45. Colloquium Mosbach
The Cytoskeleton

45. Colloquium der Gesellschaft für Biologische Chemie
14.-16. April 1994 in Mosbach/Baden

The Cytoskeleton

Edited by B. M. Jockusch, E. Mandelkow
and K. Weber

With 64 Figures

Springer

Professor Dr. B. M. JOCKUSCH
Technische Universität Braunschweig
Zoologisches Institut
Abteilung Zellbiologie
Spielmannstr. 7
D-38092 Braunschweig
Germany

Professor Dr. E. MANDELKOW
Max-Planck-Gesellschaft
AG für strukturelle Molekularbiologie
Notkestr. 85
D-22607 Hamburg
Germany

Professor Dr. K. WEBER
Max-Planck-Institut für Biophysikalische Chemie
Am Faßberg
D-37018 Göttingen
Germany

ISBN-13:978-3-642-79484-1 e-ISBN-13:978-3-642-79482-7
DOI: 10.1007/978-3-642-79482-7

Library of Congress Cataloging-in-Publication Data. Gesellschaft für Biologische Chemie. Colloquium (45th : 1994 : Mosbach, Baden-Wüttemberg, Germany) The cytoskeleton / 45. Colloquium der Gesellschaft für Biologische Chemie 14.-16. April 1994 in Mosbach/Baden; edited by B.M. Jockusch, E. Mandelkow, and K. Weber. p. cm. Includes bibliographical references.ISBN-13:978-3-642-79484-1 (acid-free paper) 1. Cytoskeleton–Congresses. I. Jockusch, B.M. (Brigitte M.), 1939-. II. Mandelkow, E. (Eckhard), 1943- . III. Weber,Klaus, Dr., Prof. IV. Title. QH603.C96G47 1994 574.87'34–dc20 94-47371

Softcover reprint of the hardcover 1st edition 1995

Cover design: Bruno Winkler Heidelberg
Typesetting: M. Masson-Scheurer, Homburg/Saar
SPIN 10126654 39/3130-5 4 3 2 1 0 - Printed on acid-free paper

Preface

This volume contains the proceedings of the 45th Mosbach Colloquium of the German Society for Biological Chemistry (GBCh). The 1994 meeting was the first in this series devoted to the cytoskeleton. This complex system enables the eukaryotic cell to form discrete contacts with neighboring cells and the extracellular matrix, to differentiate, to move, change shape, transport organelles, and proliferate. These diverse tasks are performed by three distinct fibrillar networks: microfilaments, microtubules, and intermediate filaments, which are composed of structural and regulatory elements. The precise interplay between the components in time and space determines which of the various functions is performed. The rapid progress made in this field is best exemplified by the recent unraveling of the molecular mechanism of intracellular movement. Here, the modern microscopies of today allow the motility to be visualized in realtime. Molecular biology has dissected the functional domains of the motor proteins involved and provided material required for biochemical studies as well as structure analysis by X-ray diffraction and NMR. Thus, it is now possible to integrate cellular behavior and molecular structure in a unifying picture. Other recent advances in the field point to the role of cytoskeletol proteins in human diseases, from allergies to skin blistering, atrophies, and Alzheimers's disease. The book is an up-to-date account of the field told by an international set of experts, a broad introduction to newcomers, and a valuable reference for practitioners.

January 1995

Brigitte M. Jockusch, Braunschweig
Eckhard Mandelkow, Hamburg
Klaus Weber, Göttingen

Contents

Contributors

You will find the addresses at the beginning of the respective contribution

Microtubule Dynamics and Chromosome Movement

J. R. McIntosh, V. A. Lombillo, C. Nislow, and E. A. Vaisberg[1]

1 Introduction

The mitotic spindle is responsible for organized chromosome movement during cell division. The spindle is composed largely of microtubules (MTs) and their associated proteins, but the mechanisms by which these components interact to produce mitosis with a low frequency of error in chromosome segregation is still a subject of intensive investigation.

Studies in the 1950s and 60s demonstrated that the spindle is a dynamic assembly of protein polymers that forms as the chromosomes became organized and disassembles as they are segregated (reviewed in Bajer and Mole-Bajer 1972). Spindle structure at metaphase is labile, showing sensitivity to subunit dilution, reduced temperatures, high hydrostatic pressure, or to drugs that bind tubulin, the principal subunit of a MT (reviewed in Inoue 1981). Gradual disassembly of spindle Mts by experimental perturbation allowed the chromosomes and poles to approach one another, much as they do in normal anaphase, leading several workers to propose that it is the assembly and disassembly of MTs themselves that drives chromosome movement (Inoue, 1981).

A competing view of mitotic mechanism has been developed through analogies with two well-studied forms of biological motility: muscle (e.g., Forer 1974, who has proposed that actin and/or myosin play mechanochemical roles in mitosis), and cilia or flagella, (e.g., McIntosh et al. 1969, who proposed that ATPases like dynein might induce the relative movement of spindle MTs). Opinions about mitotic mechanism have recently swayed strongly towards models based on motor enzymes, due largely to the discovery in spindles of both a cytoplasmic isozyme of dynein and several kinesin-like, MT-dependent motor enzymes (reviewed in McIntosh and Pfarr 1991; Sawin and Scholey 1991). It is now clear that a spindle contains many motor enzymes (reviewed in Goldstein 1993), and many investigators would hold that all mitotic chromosome movement is dependent upon these ATPases.

Such clarity of opinion has recently been clouded by the demonstration that MT disassembly can do work on objects that remain attached as the polymers lose subunits (Koshland et al. 1988; Coue et al. 1991). It is now certain that tubulin disassembly in the absence of sufficient ATP to activate motor enzymes can exert forces ≥1 pN on a chromosome that remains attached to the MT. The velocities of such movements (ca. 25 μm/min) are within the range of normal physiological chromo-

[1] Department of Molecular, Cellular and Developmental Biology, University of Colorado, Boulder 80309-0347, USA.

45. Colloquium Mosbach 1994
The Cytoskeleton

some motions (1–50 μm/min, depending on the stage of mitosis) and vesicular traffic in cells (ca. 60 μm/min). Disassembly-dependent motility occurs as subunits are lost from the MT plus ends, so it is analogous to the motions that occur during anaphase A, when chromosomes approach the poles. Such observations have reactivated interest in the possibility that MT disassembly may contribute to the mitotic motions of chromosomes.

In the work described below, we review some of the recent studies from our laboratory in which movements that occur with MT disassembly have been probed with antibodies to identify the motor molecules that may be important for holding chromosomes onto disassembling MTs. We then describe a simpler system in which microspheres coupled to purified proteins are bound to MTs and moved by tubulin disassembly. The implications of these data for mechanisms of chromosome movement in cells are then discussed.

2 Materials and Methods

The methods for obtaining MTs of known orientation that are tethered to a glass substrate are described in Lombillo et al. (1993). Methods for treating chromosomes with the antibodies used to perturb in vitro chromosome movement are described in Lombillo et al. (1995a). The methods for coating microspheres with protein and following their movement are described in Lombillo et al. (1995b).

3 Results

3.1 The Attachment of Chromosomes near MT Plus Ends and Disassembly-Dependent Chromosome Movement

Our system for generating MTs of known orientation that are tethered to a glass coverslip by their minus end is depicted in Fig. 1. The basal bodies resident in exhaustively lysed *Tetrahymena* are still active in nucleating tubulin polymerization, so the "pellicle" that is left after detergent extraction and rinsing can be attached to a glass coverslip and assembled into a flow cell suitable for light microscopy. Tubulin is then introduced into the chamber under conditions that will promote its assembly to form a forest of MTs with their plus ends distal to their sites of attachment to the coverslip, as diagrammed in Fig. 1.

Chromosomes or other particles may then be introduced into the chamber, and so long as the conditions continue to favor tubulin assembly, the MTs growing from the pellicle persist. Once chromosomes have become attached to the MTs, the preparation can be rinsed with protein-free tubulin assembly buffer (0.1 M PIPES, pH 6.9, 2 mM $MgCl_2$, 1 mM EGTA, and 2 mM DTT), and the MTs begin to dissolve, even at 35 °C. MTs with a free plus end dissolve first, but those with a chromosome attached eventually follow suit. As the MTs shorten, approximately half the chromosomes remain

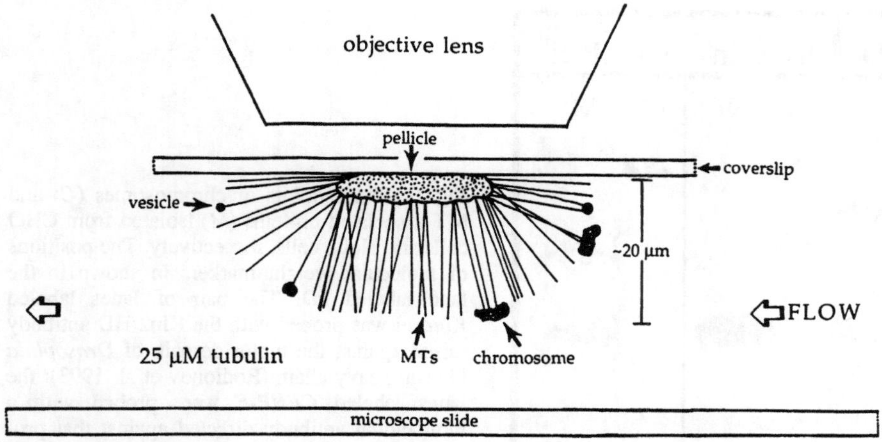

Fig. 1. A diagram of the perfusion chamber used in these experiments

attached to the shrinking polymers and move in towards the pellicle in association with subunit loss (Coue et al. 1991). The other chromosomes detach and wash away in the buffer flow, presumably because either the strength of their attachment to the MT was not sufficient to withstand the forces generated by the viscous drag of the flowing buffer or the activity of their kinetochores was not preserved.

3.2 An Immunological Approach to the Identification of Kinetochore Proteins Involved in Disassembly-Dependent Movement of Chromosomes

To identify proteins that may contribute to this phenomenon, we have applied antibodies to the chromosomes and asked whether they affect either chromosome attachment to the MT arrays or the movements that occur upon tubulin dilution. Since dynein is localized at kinetochores (Pfarr et al. 1990; Steuer et al. 1990) and is a minus end-directed motor enzyme, we have applied antibodies raised against recombinant pieces of cytoplasmic dynein that include its putative active site (Vaisberg et al. 1993). These antibodies block the ATP-dependent movement of MTs caused by dynein bound to glass coverslips, but they have no effect upon chromosome movement in our in vitro assay. They also fail to block chromosome movement in vivo (Vaisberg et al. 1993).

Kinesin-like proteins have also been localized to kinetochores (Yen et al. 1992; Wordeman and Mitchison 1994). We therefore treated chromosomes with a broadly cross-reacting antibody raised against a recombinant fragment of the motor domain of *Drosophila* kinesin, which has been demonstrated to block kinesin-dependent movements both in vitro and in vivo (Gyoeva and Gelfand 1991; Rodionov et al. 1993). Application of this antibody at concentrations that stop ATP-dependent movement of MTs over kinesin attached to glass do not affect the binding of the chromosomes to MTs, but they completely inhibit chromosome movement in our assay. This result

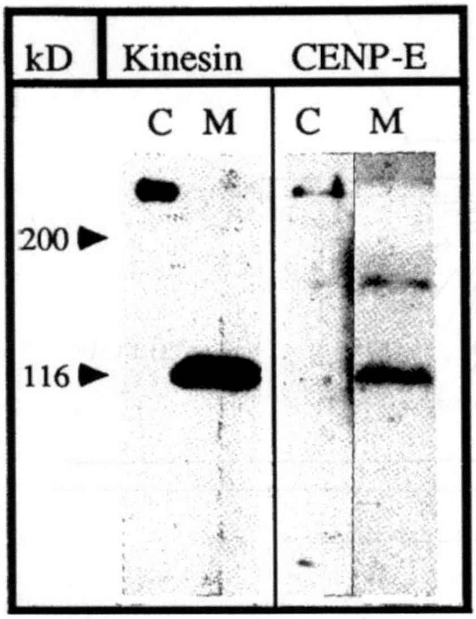

Fig. 2. Immunoblots of chromosomes (*C*) and MT associated proteins (*M*) isolated from CHO cells and HeLa cells, respectively. The positions of molecular weight markers in shown in the lane labeled *kD*. The pair of lanes labeled *Kinesin* was probed with the Kin2/HD antibody raised against the motor domain of *Drosophila* kinesin heavy chain (Rodionov et al. 1993); the lanes labeled *CENP-E* were probed with a monoclonal antibody directed against that protein (Yen et al. 1992). The component with an apparent molecular weight of *116 kD* is kinesin heavy chain, and the band at ca. *250 kD* is CENP-E

implicates a kinesin-like protein as an important component of the system that attaches chromosomes to disassembling MTs.

To identify the kinetochore components that react with this antibody, we have performed immunoblots on purified chromosomes. Only one electrophoretic component is recognized by this broadly cross-reacting, function-blocking antibody (Fig. 2). Rescreening of the same blot with specific antibodies against particular kinesin-like proteins has revealed that this band is CENP-E, the high molecular weight, kinesin-like protein known to associate with mammalian kinetochores (Yen et al. 1992). It therefore seemed plausible that CENP-E is involved in the minus end-directed movement brought about by tubulin disassembly in vitro (Lombillo et al. 1995).

To test this inference, we have collaborated with Dr. Timothy Yen of the Fox Chase Cancer Research Institute, Philadelphia, PA, and applied a series of antibodies raised against recombinant fragments of CENP-E to chromosomes prior to using them in our assay. Yen has expressed parts of CENP-E in bacteria: its kinesin-like head domain (the part that probably serves as a motor), the tail, and the neck region that connects head and tail. These polypeptides have been used to immunize rabbits and the resulting IgG fractions purified on a Protein-A columns (Lombillo et al. 1995). Both the tail and the head antibodies reduced the velocity of chromosome movement by a factor of about three, while the neck antibodies blocked motion completely. The latter are as effective in stopping chromosome movement as the more broadly cross-reacting antibodies raised against the head domain of *Drosophila* kinesin. We conclude that CENP-E probably plays an important role in coupling chromosomes to MTs so that tubulin depolymerization can occur while the chromosome remains bound.

3.3 Reconstitution of Disassembly-Dependent Movement from Simpler Components

The biochemical complexities of chromosomes and kinetochores raise concerns that antibodies to one kinetochore component might indirectly affect the behavior of neighboring molecules. Thus, there is reason to worry that the antibodies to CENP-E might be acting indirectly on the molecules that are responsible for disassembly-dependent chromosome movement. We have therefore sought a better-defined system in which to study the molecular players in disassembly-dependent motility. To this end, we have coated latex microspheres with different motor molecules or MT-associated proteins, introduced them into our chamber, tested their binding to MTs, and examined their ability to effect microsphere movement when MT disassembly is induced.

Initial control preparations included latex spheres, both charged and uncharged, with no added protein molecules. These failed to bind to MTs in our assay. We have coated beads with tubulin or bovine serum albumin and obtained the same result. When beads are coated with a preparation of crude MT-associated proteins from bovine brain, they stick well to the pellicle-initiated MTs, but during MT disassembly they do not move; they simply wash away in the buffer flow.

Flagellar dynein has been shown to bind MTs in the presence of MgATP and vanadate ions, adhering to the MT wall in such a way that there is no directed movement, but the tethered polymer shows one-dimensional diffusion along its axis (Vale et al. 1989). Vale and his collaborators have interpreted this result to suggest that the ternary complex of dynein with ADP and VO_4^{-3} binds weakly to MTs, tethering them to the glass, but permitting their motion under the action of simply thermal energy. This result seemed sufficiently similar to the results we were obtaining with chromosomes in vitro that flagellar dynein was our starting point in a quest for motor enzymes that would support bead movement in vitro. When dynein from the flagella of sea urchin sperm is adsorbed to beads and then introduced into the chamber in the buffer described by Vale (10 mM Tris-HCl, pH 7.5, 50 mM KCl, 1 mM $MgCl_2$, and 1 mM EGTA with ATP and $NaVO_4$), the dynein-coated beads attach to the MTs, and upon solubilization of the MTs, about 1/5th of the beads move.

Similar preparations of beads coated with kinesin prepared from HeLa cells by the method of Hollenbeck et al. (1989) will move under similar conditions, but with much better efficiency; more than half of the beads move upon MT disassembly with 1 mM $CaCl_2$ (Fig. 3). Interestingly, kinesin-coated beads bound to MTs in the presence of conventional MT assembly buffer do not move upon disassembly, but simply wash away in the buffer flow. We conclude that the buffer conditions are important for the way in which a motor enzyme will bind to MTs, and that, in the absence of ATP, both dynein and kinesin can support movement dependent on MT disassembly.

We were curious to see if the cytoplasmic isozyme of dynein, the form localized to kinetochores, could also support MT disassembly-dependent movement in vitro. All our efforts to produce such movement have failed. These included the use of both tubulin assembly buffer, the Vale 1-D diffusion buffer, and a series of modifications of the latter made by increasing salt in small steps, trying to reduce the avidity of dynein-MT binding. The role that dynein may play in kinetochore attachment to MTs is thus far unclarified by our work.

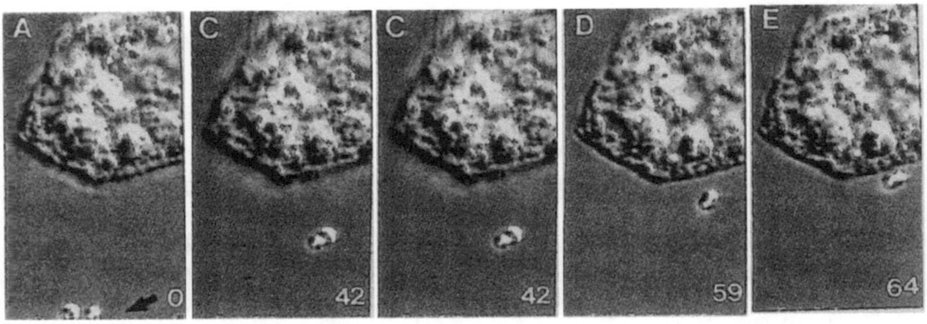

Fig. 3 A–E. A series of video images obtained with the chamber shown in Fig. 1, using differential interference contrast optics to image a pellicle (the polygonal structure) and two latex microspheres coated with kinesin isolated from HeLa cells. The *arrow* in **A** shows the direction of buffer flow. Times are given in seconds following tubulin dilution. The MTs that connect the pellicle with the microspheres are not visible in this printing

Kinesine-like molecules have been identified that move along MTs in either the plus or the minus directions (reviewed in Goldstein 1993). Several labs are working to identify the portions of the motor enzyme's head domain that defines its direction of motility. During work of this kind, Russell Stewart discovered a chimeric gene, constructed from the DNA encoding *Drosophila* kinesin's head domain and the tail from the kinesin-like protein, NCD. The resulting protein binds to MTs and displays 1-D diffusion, but fails to promote conventional ATP-dependent movement (Stewart et al. 1993). This chimera, called NK350, has been tested in our assay. It supports the most rapid movement of any disassembly-dependent process we have observed. In standard MT assembly buffer, microspheres move 186 ± 84 µm/min ($N = 45$). This velocity not only demonstrates that NK350 is effective at coupling MTs with other objects so that movement is allowed, it also suggests that the coupling factor can have an effect on the tubulin disassembly process itself. The rate of MT rapid shortening in a conventional tubulin assembly buffer is ca. 25 µm/min (Walker et al. 1988; Lombillo et al. 1995). This speed is normally seen when tubulin subunits are falling off after extensive subunit dilution, so it is thought to represent the maximum velocity of MT disassembly in this buffer. The mean speed of beads coated with NK350 in the same buffer is ca. sevenfold higher, suggesting that the chimera bound to the bead is interacting with the end of the MT in such a way as to catalyze subunit loss. We have tested the possibility that soluble NK350 is a MT disassembly factor, but in solution no such effect is seen. We conclude that only a structure-bound motor enzyme with the right sort of MT binding domain can facilitate subunit loss from the end of a disassembling MT.

4 Discussion

Our results from antibody treatments of chromosomes suggest that kinesin-like molecules associated with the kinetochore play a role in coupling chromosomes to MTs so that tubulin depolymerization can affect chromosome movement. Our results with protein-coated beads in vitro demonstrate that kinesin-like proteins can fulfill the same role in simpler systems. Moreover, the movements are accomplished by an association of the MT-binding protein with the disassembling MT end that can affect the rate of tubulin disassembly. We infer that there is a direct interaction between the motor domain of the microsphere-bound motor enzymes and the tubulin subunits near the MT end such that tubulin disassembly is promoted.

Images of disassembling MTs obtained by cryomicroscopy have revealed protofilaments that bend outward from the MT axis (Mandelkow et al. 1991). A plausible mechanism by which motor enzymes might affect MT disassembly is through encouraging this lattice distortion and thus the rate of subunit loss from the disassembling end (Fig. 4). Alternatively, the motor enzyme might alter the GTPase activity or GDP binding of the terminal MT subunits, thereby changing the rates at which they depolymerize (Davis et al. 1994).

While our work demonstrates that MT disassembly can do work on chromosomes, it is not obvious that this process plays a role during chromosome movement in vivo. Clearly there are motor enzymes in spindles, not only at kinetochores, but also in the environs of the spindle poles and in the regions between (reviewed in McIntosh and Pfarr 1991; Sawin and Scholey 1991). Indeed, at least four kinesin-like proteins contribute to spindle function in the yeast *Saccharomyces cerevisiae* (Meluh et al. 1990; Hoyt et al. 1992; reviewed in Page and Snyder 1993), and it seems likely that even greater numbers of enzymes will contribute to mitosis in higher eukaryotic cells. What mitotic role, then, can one expect for tubulin disassembly?

It is evident that spindle MTs change their lengths during mitosis. During both prometaphase, when chromosomes attached to the spindle are migrating towards the equator, and anaphase A, when the chromosomes approach the poles, the kinetochore-associated MTs change length. These length changes are important because prometaphase motion to the spindle equator is probably significant for the accuracy of chromosome segregation, and the anaphase approach to the spindle pole increases the separation between sister chromosomes before cytokinesis. Moreover, the framework

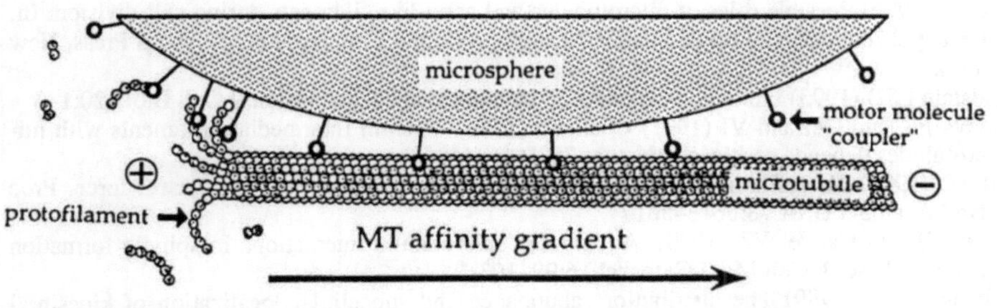

Fig. 4. Model for the events that couple motor movement to MT disassembly

of MTs that connects the spindle poles through much of mitosis elongates during anaphase B. This elongation results from MT polymerization, combined with sliding that is almost certainly dependent upon MT motor enzymes (Hogan and Cande 1990; Nislow et al. 1992, reviewed in McIntosh 1994). In all these cases, MT dynamics is combined with MT-dependent motor enzyme action to contribute to chromosome segregation in mitosis. It is therefore clear that MT polymerization and depolymerization, like motor enzyme action, are intrinsic to spindle action.

Our work demonstrates that MT disassembly can do work, and it follows from the symmetry of the equations that describe the free energy changes associated with the depolymerization process that forces acting on MTs may also affect MT polymerization (Hill 1981). We interpret our observations to suggest that one is now seeing parts of a system of linked reactions by which tubulin polymerization and depolymerization may be coupled through motor enzyme action to mechanical events of significance to the cell. We expect that MT depolymerization will affect the action of motor enzymes and that motor enzyme action will affect the extent and rate of tubulin assembly. In short, we expect that the MT elongations and shortenings that occur in the course of a normal mitosis are often coupled with motor enzyme action. We suggest that the ATP-dependent and ATP-independent actions of motors work together with tubulin assembly dynamics to generate and regulate the forces that push and pull on chromosomes and achieve their segregation with high fidelity.

Acknowledgments. The DNA sequences essential for our work with antidynein were obtained by Dr. Michael Koonce and Paula Grissom. The antibodies essential for the studies inhibiting chromosome motion in vitro were contributed by Professor Vladimir Gelfand of the University of Illinois (Antibody to *Drosophila* Kinesin Motor Domain) and Dr. Timothy Yen of Fox Chace Cancer Research Center (Antibodies to CENP-E). We are very grateful for all this help. This work was supported in part by grants from the NIH to J.R.M. (GM33787 and GM36663).

References

Bajer AS and Mole-Bajer J (1972) Spindle dynamics and chromosome movements. Int Rev Cytol Suppl 3:1–271

Coue M, Lombillo VA and McIntosh JR (1991) Microtubule depolymerization promotes particle and chromosome movement in vitro. J Cell Biol 112:1165–1175

Davis A, Sage CR, Dougherty CA and Farrell KW (1994) Microtubule dynamics modulated by guanosine triphosphate hydrolysis activity of beta-tubulin. Science 264:839–842

Forer A (1974) Possible roles of microtubules and actin-like filaments during cell-division. In: Cell cycle controls. Ed by Padilla GM, Cameron IL and Zimmerman AM. Acad Press, New York

Goldstein LSB (1993) Functional redundancy in mitotic force generation. J Cell Biol 120:1–3

Gyoeva FK and Gelfand VI (1991) Coalignment of vimentin intermediate filaments with microtubules depends on kinesin. Nature 353:445–448

Hill TL (1981) Microfilament or microtubule assembly or disassembly against a force. Proc Natl Acad Sci USA 78:5613–5617

Hogan CJ and Cande WZ (1990). Antiparallel microtubule interactions in spindle formation and anaphase B. Cell Motil Cytoskel 16:99–103

Hollenbeck PJ (1989) The distribution, abundance and subcellular localization of kinesin. J Cell Biol 108:2335–2342

Hoyt MA, He L, Loo KK and Saunders WS (1992) Two *S cerevisiae* kinesin-related gene products required for mitotic spindle assembly. J Cell Biol 118:109–120

Inoue S (1981) Cell division and the mitotic spindle. J Cell Biol 91:131s–147s

Koshland DE, Mitchison TJ and Kirschner MW (1988) Polewards chromosome movement driven by microtubule depolymerization in vitro. Nature 331:499–504

Lombillo VA, Coue M and McIntosh JR (1993) In vitro motility assays using microtubules tethered to tetrahymena pellicles. Meth Cell Biol 39:148–165

Lombillo VA, Nislow C, Yen TJ, Gelfand VI and McIntosh JR (1995a) Antibodies to the kinesin motor domain and CENP-E inhibit microtubule depolymerization-dependent motion of chromosomes in vitro. J Cell Biol 128:107–115

Lombillo VA, Stewart RJ and McIntosh JR (1995b) Minus-end-directed motion of kinesin-coated microspheres driven by microtubule depolymerization. Nature 373:161–164

Mandelkow EM, Milligan R and Mandelkow E (1991) Microtubule dynamics and microtubule caps: a time-resolved cryo-electron microscope study. J Cell Biol 114:977–991

McIntosh JR (1994) In: Microtubules. Ed by Hyams JS and Lloyd CW. Wiley-Liss, Inc New York pp 413–434

McIntosh JR, Hepler PK and Van Wie DG (1969) Model for mitosis. Nature 224:659–663

McIntosh JR and Pfarr CM (1991) Mitotic motors. J Cell Biol 115:577–585

Meluh PB and Rose MD (1990) Kar3, a kinesin-related gene required for yeast nuclear fusion. Cell 60:1029–1041

Nislow C, Lombillo VA, Kuriyama R and McIntosh JR (1992) A plus-end-directed motor enzyme that moves antiparallel microtubules in vitro localizes to the interzone of mitotic spindles. Nature 359:543–547

Page BD and Snyder M (1993) Chromosome segregation in yeast. Annu Rev Microbiol 47:231–261

Pfarr CM, Coue M, Grissom PM, Hays TS, Porter ME and McIntosh JR (1990) Cytoplasmic dynein is localized to kinetochores during mitosis. Nature 345:263–265

Rodionov VI, Gyoeva FJ, Tanaka E, Bershadsky AD, Vasiliev JM and Gelfand VI (1993) Microtubule-dependent control of cell shape and pseudopodial activity is inhibited by the antibody to kinesin motor domain. J Cell Biol 123:1811–1820

Sawin K and Scholey JM (1991) Motor proteins in cell division. Trends in Cell Biol 1:122–129

Steuer ER, Wordeman L, Schroer TA and Sheetz MP (1990) Localization of cytoplasmic dynein to mitotic spindles and kinetochores. Nature (London) 345:266–268

Stewart RJ, Thaler JP and Goldstein LSB (1993) Direction of microtubule movement is an intrinsic property of the motor domains of kinesin heavy chain and *Drosophila* ncd protein. Proc Natl Acad Sci USA 90:5209–5213

Vaisberg G, Koonce MP and McIntosh JR (1993) Cytoplasmic dynein plays a role in mammalian mitotic spindle formation. J Cell Biol 123:849–858

Vale RD, Soll DR and Gibbons IR (1989) One dimensional diffusion of microtubules bound to flagellar dynein. Cell 59:915–925

Walker RA, O'Brien ET, Pryer NK, Soboeiro MF, Voter WA, Erickson HP, Salmon ED (1988) Dynamic instability of individual microtubules analyzed by video light microscopy: rate constants and transition frequencies. J Cell Biol 107:1437–1448

Wordeman LG and Mitchison TJ (1994) Identification and characterization of mitotic centromere-associated kinesin, that associates with centromeres during mitosis. J Cell Biol 128:95–106

Yen TJ, Li G, Schaar BT, Szilak I and Cleveland DW (1992) CENP-E is a putative kinetochore motor that accumulates just before mitosis. Nature 359:536–539

New Motilities and Motors in the Flagella of *Chlamydomonas*

M. BERNSTEIN[1], P. L. BEECH[1], K. A. JOHNSON[2], K. G. KOZMINSKI[1], and J. L. ROSENBAUM[1]

Chlamydomonas is a haploid, unicellular green alga, about 10 μm in diameter. Synchronous cultures can be grown phototrophically in a simple defined medium. In addition to the two 10–12 μm-long flagella that emanate from one end of the cell, *Chlamydomonas* shares the cytology of typical higher plant and animal cells (e.g., basal bodies/centrioles, microtubules, actin filaments, chloroplasts, mitochondria, endoplasmic reticulum, and Golgi). The genetics of *Chlamydomonas* is similar to that of yeast, with plus and minus cells mating to form zygotes, followed by meiosis and tetrad formation. Many mutants are available (Harris 1989) or can be generated by classical mutagenesis procedures. Because *Chlamydomonas* is easily transformed, with the introduced DNA inserting randomly into the genome (Diener et al. 1990; Kindle et al. 1989), mutants can also be generated by insertional mutagenesis, allowing for the isolation of mutant genes by plasmid rescue (Tam and Lefebvre 1993). In addition, progress is being made on gene cloning by complementation (Purton and Rochaix 1994) and on the selection of homologous recombinants (Sodeinde and Kindle 1993). We are using flagellar regeneration in this green alga as a model system for studying the biogenesis of cell organelles (Johnson and Rosenbaum 1993).

The flagella of *Chlamydomonas* can be detached by several physical and chemical procedures; new flagella regenerate in about 90 min. Flagellar detachment induces the synthesis of flagellar protein mRNAs (e.g., tubulins, dyneins, radial spokes) (Baker et al. 1984; Schloss et al. 1984; Silflow and Rosenbaum 1981; Fig. 1) and the accumulation of flagellar proteins in the cytoplasm (Lefebvre and Rosenbaum 1986; Remillard and Witman 1982). These flagellar proteins rapidly find their way to the basal bodies at one end of the cell and begin to assemble new flagella within minutes. One can observe short, beating flagella in less than 10 min. We and others have cloned and sequenced several of the genes for flagellar proteins (Curry et al. 1992; Silflow et al. 1985; Williams et al. 1989; Youngblom et al. 1984) and, by transformation of cells mutant in these genes, have begun to study several problems concerning the synthesis and assembly of flagellar proteins (Diener and Ang 1993; Diener et al. 1990). One question is the site of flagellar protein synthesis: are flagellar proteins synthesized throughout the cytoplasm and then transported to the basal bodies, or are they initially synthesized close to the basal bodies? We would also like to know whether the over 150 different flagellar proteins assemble separately onto the microtubules as they elongate or if they first form cytoplasmic complexes which then assemble onto the growing flagellum. Do the polypeptides which compose the dynein arms, for exam-

[1] Yale University, Department of Biology, Kline Biology Tower, P.O. Box 208103, New Haven, CT, 06520-8103, USA.
[2] Department of Biology, Haverford College, Haverford, PA, 19041, USA.

45. Colloquium Mosbach 1994
The Cytoskeleton

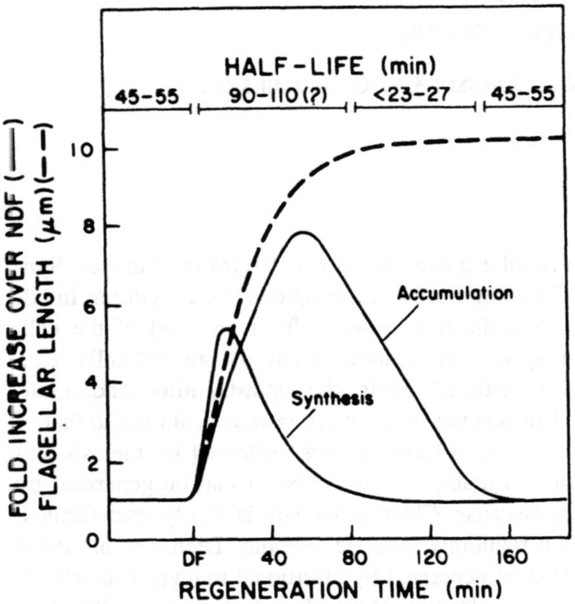

Fig. 1. Diagram summarizing changes in tubulin mRNA synthesis, accumulation and half-life during flagellar regeneration. (Baker et al. 1984)

ple, first assemble into complexes before entering the flagellum? As a related question: how do the flagellar proteins find their way into the flagellar compartment? Are they targeted by specific sequences on the flagellar proteins? Here, we will discuss the problem of how flagellar proteins, once they enter the flagellar compartment, arrive at their assembly site at the flagellar tip (Johnson and Rosenbaum 1992; Rosenbaum et al. 1969).

To determine the polarity of assembly of *specific* flagellar proteins once they enter the flagellum, we used the ability of *Chlamydomonas* to form quadriflagellate dikaryons by the fusion of cells during mating. *Chlamydomonas* cells expressing an epitope-tagged α-tubulin construct were fused by mating with cells that had half-length regenerating flagella. Following fusion, the pair of half-length flagella of each quadriflagellate elongated to full-length by assembling the epitope-tagged tubulin supplied by the donor cell (Fig. 2). Immunofluorescence microscopy, using an antibody to the epitope tag, showed that all of the epitope-tagged tubulin had added onto the distal ends of the half-length flagella (Fig. 3). This was confirmed by immunoelectron microscopy of negatively stained flagellar axonemes of the quadriflagellates. A more surprising result was obtained in similar experiments when wild-type cells with full-length flagella were fused with the paralyzed mutant, *pf14*. This mutant has full-length flagella, but lacks radial spokes because it is missing radial spoke protein-3 (rsp3), the protein which attaches the spoke to the A-tubule of the outer doublet microtubules (Diener and Ang 1993). Immunofluorescence microscopy of the resulting quadriflagellates showed that the radial spokes, using epitope-tagged rsp3 supplied by the wild-type cell, assembled first onto the distal ends of the full-length, but spokeless, axonemes (Fig. 4). Thus, even though radial spoke assembly sites were available

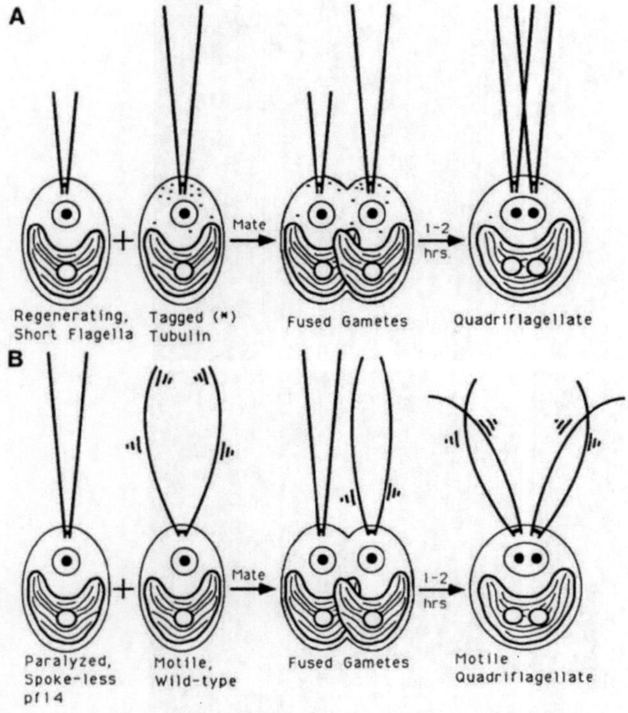

Fig. 2 A, B. A Introduction of epitope-tagged tubulin subunits into cells regenerating flagella. The recipient cells were pool-depleted gametes with one-half to two-thirds full-length flagella. The donor cells were expressing epitope-tagged α-tubulin and had full-length flagella. Within 1–2 h after mating, regeneration of the shorter pair of flagella was completed using epitope-tagged tubulin and other unassembled flagellar components from the donor cell, leaving the quadriflagellate with four equal-length flagella. These flagella were examined for the distribution of tagged tubulin. **B** Diagram of dikaryon rescue of the radial spokeless mutant *pf14*. Radial spokeless, paralyzed gametes were mated with wild-type gametes. Immediately after mating, the quadriflagellates have a pair of paralyzed flagella and a pair of motile flagella. Within 1–2 h after mating, all four flagella of the quadriflagellates begin beating. During the time course of rescue, quadriflagellate flagella were examined for the distribution of radial spoke proteins. (Johnson and Rosenbaum 1992)

along the entire length of the spokeless axonemes, radial spoke proteins were first carried to the distal tip.

The polarity studies described above suggested the existence of a mechanism that could carry precursors to the tip of the flagellum, the site of axonemal assembly. Three types of nonbeat motility, which might be involved in such transport, have been described in *Chlamydomonas* – the gliding of cells over a substrate by use of motors associated with the flagellar membrane, the movement of polystyrene beads on the external surface of the flagellar membrane, and the movement of flagellar agglutinins to the tips of the flagella during mating (Bloodgood 1992). On the basis of inhibitor experiments and ion requirements, bead movement and gliding are thought to be powered by similar motors (Bloodgood 1992). Our re-examination of these motilities

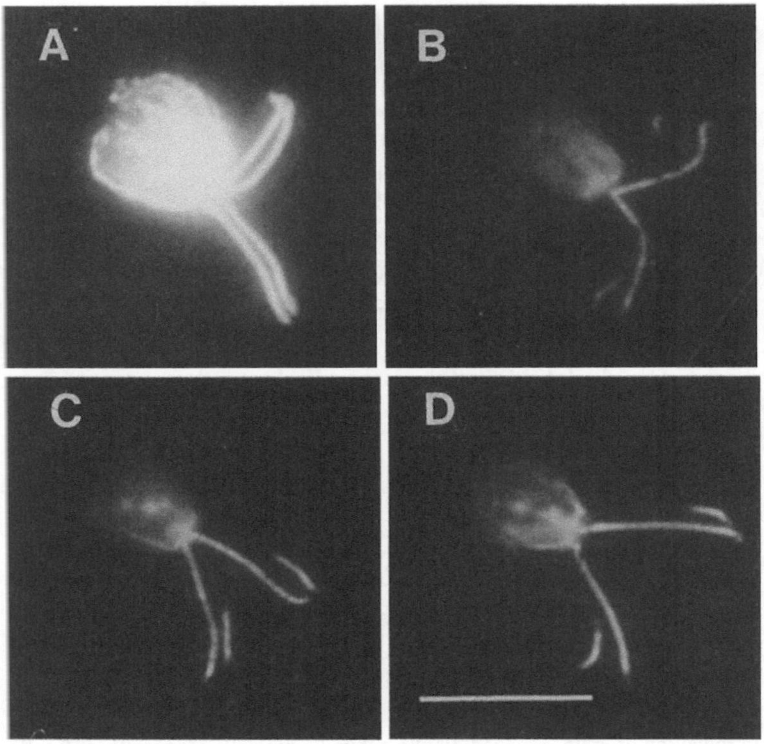

Fig. 3 A–D. Immunofluorescence localization of tubulin in quadriflagellates. **A** Using a polyclonal anti-α-tubulin antiserum, all four axonemes as well as the cytoskeletal microtubules of a dikaryon are labeled. **B–D** Using an antiepitope mAb, two of the axonemes of a quadriflagellate are labeled along their full length and two axonemes are labeled only along their distal portions. Before mating, the antiepitope antibody labels all microtubules of donor gametes (expressing the epitope-tagged tubulin) but does not label any structures in wild-type recipient gametes (data not shown). *Bar* 10 μm. (Johnson and Rosenbaum 1992)

by high-resolution video-enhanced DIC microscopy has revealed a previously unobserved flagellar motility, the rapid bidirectional movement of granule-like particles along the entire length of the flagellum (Kozminski et al. 1993) (Fig. 5). The exact nature of what is moving in this motility, termed *i*ntra*f*lagellar *t*ransport (IFT), cannot be determined by light microscopy. The movement beneath the flagellar membrane occurs at 2.0 μm/s toward the flagellar tip and 3.5 μm/s toward the base. IFT continues under a variety of experimental conditions that inhibit both bead movement on the flagellar surface and the gliding movements mediated by the flagella; it also occurs in mutants that cannot glide or move beads, suggesting a different motor mechanism (Kozminski et al. 1993). It is quite clear that the dynein associated with flagellar beating is unrelated to these movements because all four nonbeat motilities occur normally in mutants lacking outer and/or inner dynein arms. Motors other than flagellar dynein, therefore, must be involved in IFT and the other nonbeat motilities.

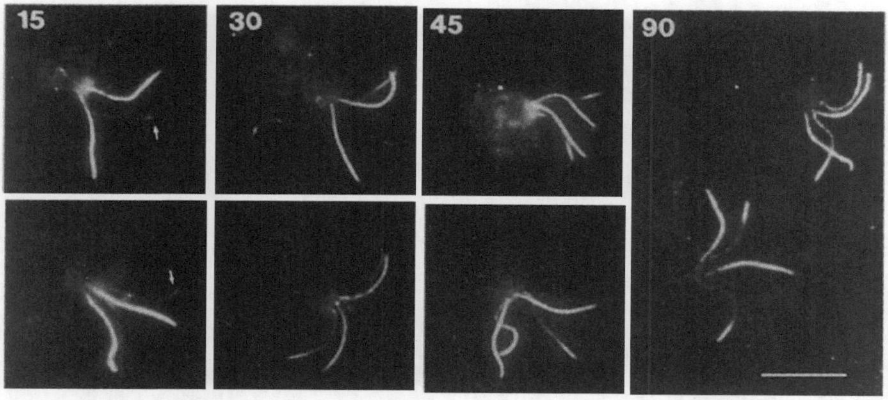

Fig. 4. Immunofluorescent localization of radial spoke protein 3 (rsp3) in quadriflagellates at various times during dikaryon rescue. Two examples are shown for each of the *15-, 30,- 45-,* and *90-min* time points. At *15 min* after mating, the two wild-type axonemes are fluorescent along their full length (they have radial spokes), whereas the two mutant axonemes are not fluorescent (they lack radial spokes and radial spoke proteins). Minor fluorescence is apparent at some of the tips of these otherwise dark axonemes (*small arrows*). By *30 min*, this fluorescence has brightened and spread towards the bases of the mutant axonemes. By *45 min*, the zone of fluorescence has expanded further. By *90 min* after mating, some of the quadriflagellates have four flagella that are fluorescent along their full lengths (upper cell of the pair). *Bar* 10 µm. (Johnson and Rosenbaum 1992)

Ultrastructural analysis has shown the presence of "rafts" containing a varying number of lollipop-shaped structures, each larger than a dynein arm, between the flagellar membrane and outer doublet microtubules (Kozminski et al. 1993). The "stick" of each lollipop is associated with the flagellar membrane and the "heads" with the B-tubule of the outer double microtubules, bridging the flagellar membrane to the microtubules (Fig. 6). These rafts, several of which can be seen in each cross or longitudinal section, may be the moving particles of IFT, although no direct evidence for this has yet been obtained.

To determine if any of the known actin or microtubule-based motors might be present in the flagella of *Chlamydomonas*, isolated flagella were analyzed on immunoblots (Fig. 7). Epitopes that reacted with antibodies to kinesin, kinesin-related proteins, cytoplasmic dynein, myosin-I, and actin were observed (Fox et al. 1994; Johnson et al. 1994; Kozminski et al. 1992). We began our molecular study of these motors by using RT-PCR to clone kinesin-like cDNAs from a *Chlamydomonas* library upregulated for flagellar mRNAs by flagellar detachment. This procedure provided evidence for three kinesin-like proteins in the flagella (Bernstein et al. 1994a). In a complementary approach, immunoblots with a variety of antibodies suggested the existence of at four flagellar kinesins (Fox et al. 1994; Johnson et al. 1994), for a total of five flageller kinesins (Bernstein and Rosenbaum 1994b). We have cloned and sequenced a full-length cDNA for one flagellar kinesin, *KLP1*, and found it to be a new member of the kinesin superfamily. An antibody raised against the nonconserved tail region of the Klp1 protein expressed in *E. coli* was used as a probe

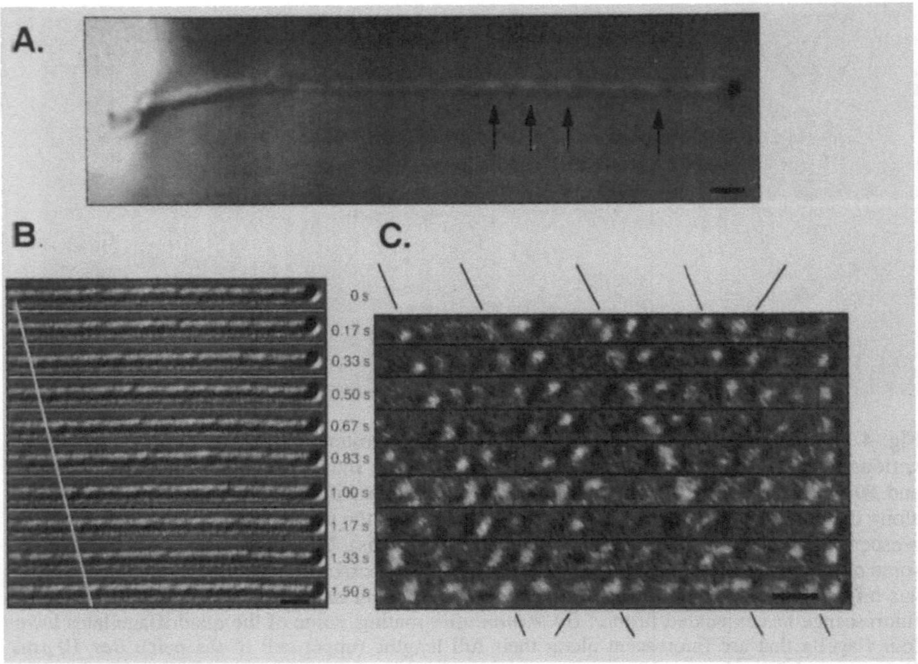

Fig. 5 A–C. Digital, video-enhanced DIC images illustrating IFT in *C. moewusii* (M475). **A** Granule-like particles can be observed as regions of high contrast along the length of a flagellum (*arrows*). The birefringent cell body is located to the left. **B** A time-lapse (seconds shown to the *right*), enhanced image sequence of a flagellum that shows the movements of a granule-like particle (*right of the white line*) toward the flagellar tip (*to the right*). Other granule-like particles (not marked) can also be seen moving toward the flagellar tip. The sampling interval between each frame in this composite image and the one shown in **C** is 0.17 s. **C** Difference-image sequence of the same flagellar region and time points shown in **B** that indicates the bidirectional movement of granule-like particles along the length of the flagellum. The vertical image height of each panel in the sequence is equal to the width of the flagellum. Only moving objects are visualized. *Bar* 1 µm. (Kozminski et al. 1993)

Fig. 6 A–F. Electron micrographs of complexes between the axonemal outer doublets and the flagellar membrane in both *C. moewusii* (**A, B**) and *C. reinhardtii* (**C–F**). In addition to these large complexes (*arrowheads*; **A, D**), small bridges are also shown (*small arrows*; **A, D**) between the flagellar membrane and the outer doublet microtubules. When viewed in longitudinal section (**B, C, E, F**), the large complexes appear to be composed of linear rafts of lollipop-shaped electron-dense projections. Panel **F** is an enlargement emphasizing the bracketed region of **E** (*mt* outer doublet microtubules; *mbr* flagellar membrane). *Bar* in **F** represents 0.23 µm for panels **A, B, D,** and **E**; 0.7 µm for **C;** 0.1 µm for **F**. (Kozminski et al. 1993)

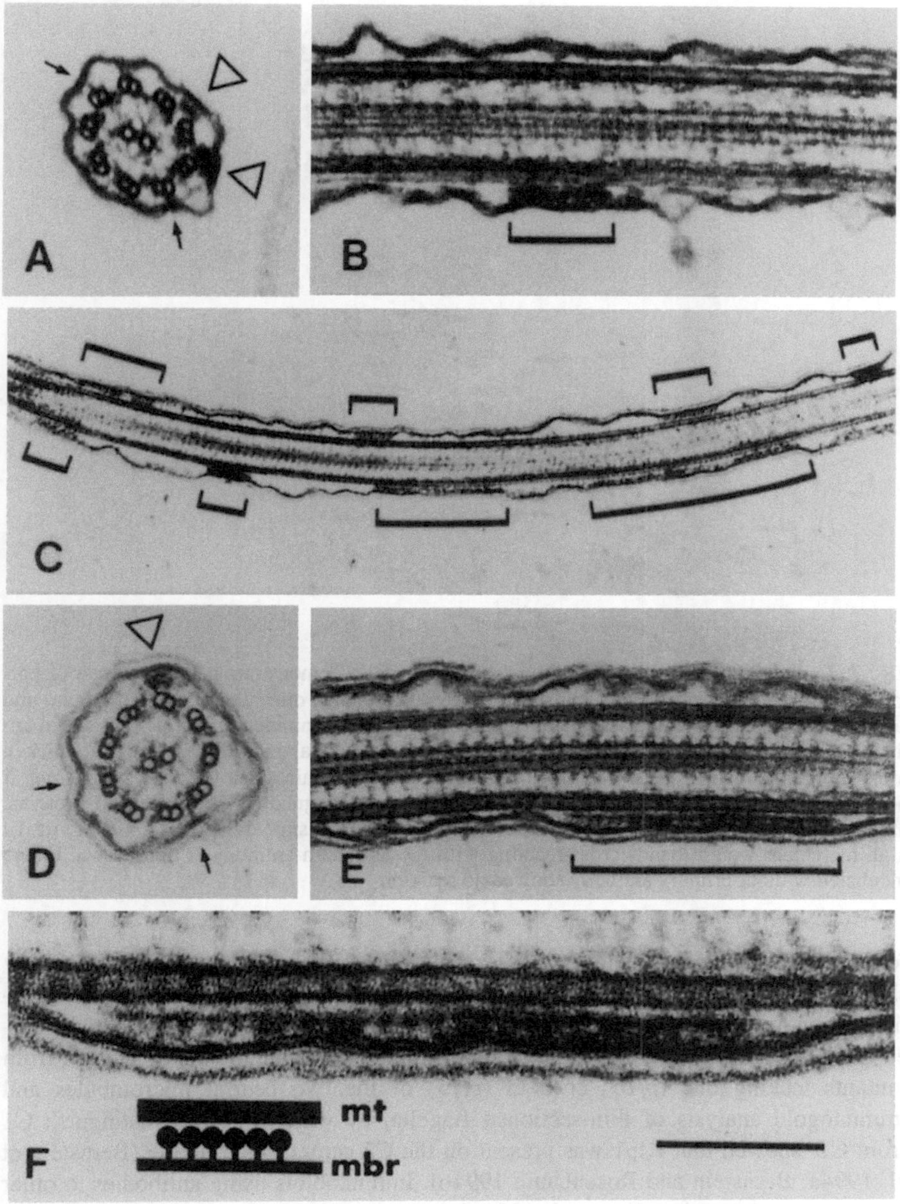
A
B
C
D
E
F
mt
mbr

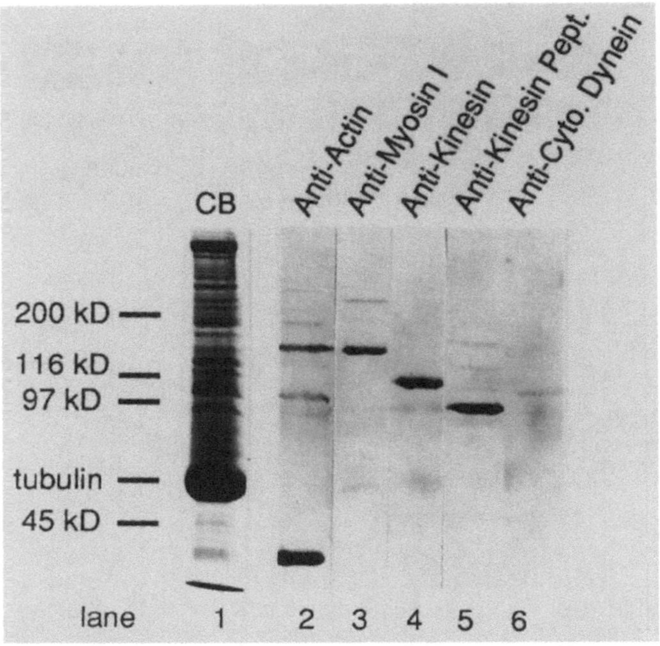

Fig. 7. Immunoblots of isolated *Chlamydomonas* flagella demonstrating the presence of epitopes recognized by antibodies against known motor proteins, other than flagellar dynein, and actin. Staining: *Lane 1* Coomassie Blue-stained SDS polyacrylamide gel; *lane 2* polyclonal antibody A5 against a *Volvox* actin decapeptide; *lane 3* polyclonal antibody R21 against chicken brush border myosin-I; *lane 4* monoclonal antibody suk4 against sea urchin kinesin; *lane 5* polyclonal antibody against conserved kinesin peptides recognizing kinesin-related proteins; *lane 6* monoclonal antibody 70.1 against chicken brain cytoplasmic dynein. The identity of the high molecular weight bands cross-reacting with the anti-actin antibody is not known. Filters incubated without primary antibody showed no reaction

for Klp1 by immunoblot and cytological analysis. Immunofluorescence microscopy showed that Klp1 was present in both cell bodies and flagella. Immunogold localization on negatively stained whole mounts of flagella showed that Klp1 was restricted to one of the two central (C) microtubules (Fig. 8). Immunoblots of flagella from mutants lacking one (*pf16*) or both (*pf15*) of the two central microtubules and immunogold analysis of thin-sectioned flagella, by which one can distinguish C1 from C2, showed that Klp1 was present on the C2 central microtubule (Bernstein et al. 1994a; Bernstein and Rosenbaum 1994b). Immunoblots using antibodies to other kinesin-related proteins (Fox et al. 1994; Johnson et al. 1994) suggested that other kinesin-related proteins are also associated with the central pair microtubules, but at this time, it is not clear which one of the two (Fox et al. 1994; Johnson et al. 1994).

Although we expected to localize a kinesin-like motor between the flagellar membrane and the axoneme, perhaps in association with the "rafts" (Fig. 6), the localization of Klp1 to the central pair microtubules shows that this kinesin is not responsible for IFT. We are proceeding, therefore, with the cloning, sequencing, and localization

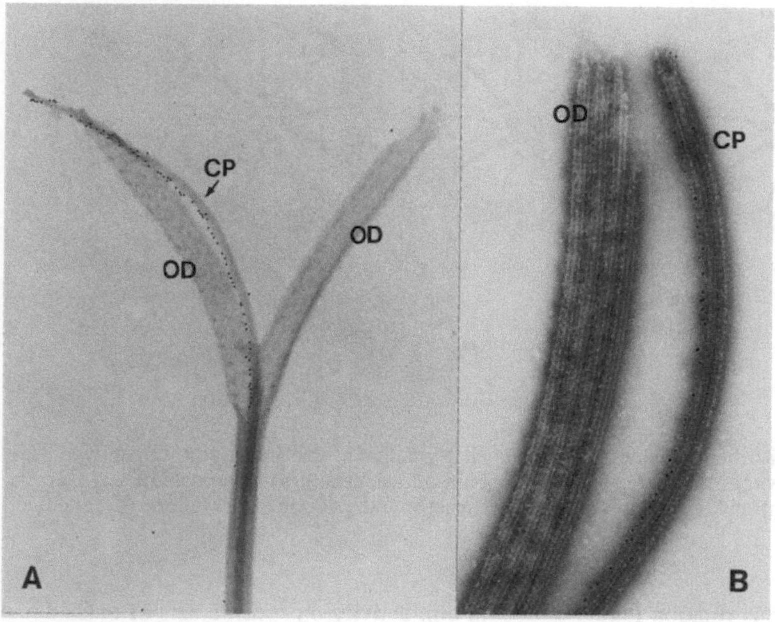

Fig. 8 A, B. Axonemes of *Chlamydomonas* prepared for wholemount transmission electron microscopy and immunolabeled with Klp1 antibody and 12-nm gold particles. **A** is a light-positive stain and **B** is a negative stain. Klp1 labeling is restricted to one side of the central pair of microtubules (*CP*) and, in the negative stain **B** it is clear that this labeling is specific to one of the microtubules only. The outer doublet microtubules (*OD*) show no Klp1 antigenicity. **A** x 21,000; **B** x 60,000

of the other flagellar kinesins, with the possibility that one of them may be involved in IFT. It is also possible that IFT is myosin-driven. Using the same RT-PCR strategy used to clone *KLP1*, we have begun the cloning of myosin cDNAs from *Chlamydomonas* (W. Mages and J. Rosenbaum, unpubl.).

Finally, the intriguing localization of Klp1 to one of the two central pair microtubules raises interesting questions both in the targeting of this kinesin and its function. To explain how this kinesin is targeted to just one out of a total of 20 flagellar microtubules (9 outer doublet + 2 central pair), it has been proposed that specific tubulin isoforms or changes in the microtubule lattice structure may serve as signals for assembly of axonemal components onto specific microtubules. However, it should be emphasized that, in *Chlamydomonas*, all microtubules are composed of just one type of α/β dimer and the tubulin lattice is probably equivalent in all microtubules. In spite of this, Klp1 can discriminate between C2 and the other 19 microtubules of the axoneme. The way in which this targeting is achieved by Klp1, or by Klp1 in association with other polypeptides of the central pair apparatus, is under investigation.

The function of Klp 1 on the central pair apparatus is equally interesting. Central pair microtubule rotation has been observed in several flagellate organisms (Kamiya 1982; Omoto and Kung 1979; Omoto and Kung 1980; Omoto and Witman 1981). It is

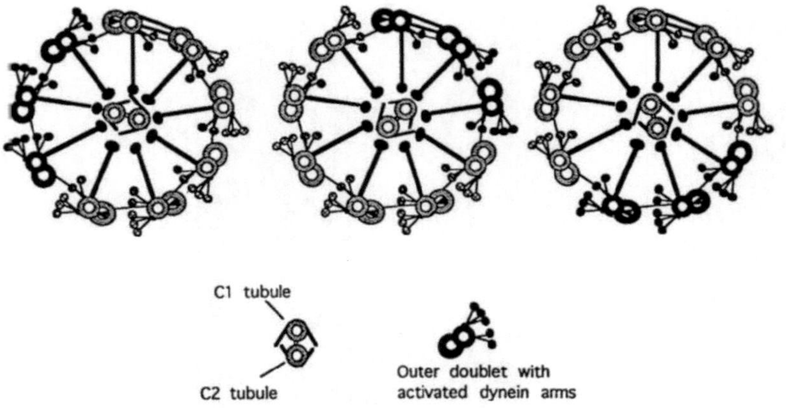

Fig. 9. The "Distributor" model for central pair regulation of dynein activity. In the hypothetical model shown, as the central pair microtubules rotate, central pair microtubule C2 interacts with three radial spoke heads at each position to activate the dynein arms on three outer doublet microtubules

thought that this rotation regulates dynein arm activity on specific sets of outer doublet microtubules, acting through the radial spokes which bridge the central pair microtubules to the outer doublets. Thus, the central pair, by rotating or twisting, may act as a kind of "distributor", allowing dynein on only certain sets of outer doublets to function (Fig. 9). If the dynein arms on all the outer doublets functioned simultaneously, no flagellar bending would occur. Considering that Klp1 may be involved in such a rotary movement, we are testing the prediction that Klp1 will rotate, rather than linearly translate microtubules, in an in vitro motility assay.

Chlamydomonas flagella, therefore, in addition to being superb organelles for studying a variety of problems of organelle assembly and dynein-based flagellar motility, also present an excellent opportunity for the application of combined biochemical, genetic, and molecular approaches to the in vivo function and properties of a number of different molecular motors, such as the kinesin-like protein, Klp1, described here.

References

Baker EJ, Schloss JA, Rosenbaum JL (1984) Rapid changes in tubulin RNA synthesis and stability induced by deflagellation in *Chlamydomonas.* J Cell Biol 99:2074–2081

Bernstein M, Beech PL, Katz SG, Rosenbaum JL (1994a) A new kinesin-like protein (Klp1) localized to a single microtubule of the *Chlamydomonas* flagellum. J Cell Biol 125:1313–1326

Bernstein M, Rosenbaum JL (1994b) Kinesin-like proteins in the flagella of *Chlamydomonas.* Trends Cell Biol 4:236–240

Bloodgood RA (1992) Directed movements of ciliary and flagellar membrane components: a review. Biol Cell. 76:291–301

Curry AM, Williams BD, Rosenbaum JL (1992) Sequence analysis reveals homology between two proteins of the flagellar radial spoke. Mol Cell Biol 12:3967–3977

Diener DR, Ang LH, Rosenbaum JL (1993) Assembly of flagellar radial spoke proteins in *Chlamydomonas*: identification of the axoneme binding domain of radial spoke protein 3. J Cell Biol 123:183–190

Diener DR, Curry AM, Johnson KA, Williams BD, Lefebvre PA, Kindle KL, Rosenbaum JL (1990) Rescue of a paralyzed-flagella mutant of *Chlamydomonas* by transformation. Proc Natl Acad Sci USA 87:5739–5744

Fox LA, Sawin KE, Sale WS (1994) Kinesin-related proteins in eukaryotic flagella. J Cell Sci 107:1545–1550

Harris EH (1989) The *Chlamydomonas* sourcebook. New York, Academic Press.

Johnson KA, Haas MA, Rosenbaum JL (1994) Localization of a kinesin-related protein to the central pair apparatus of the *Chlamydomonas reinhardtii* flagellum. J Cell Sci 107:1551–1556

Johnson KA, Rosenbaum JL (1992) Polarity of flagellar assembly in *Chlamydomonas*. J Cell Biol 119:1605–1611

Johnson KA, Rosenbaum JL (1993) Flagellar regeneration in *Chlamydomonas*: a model system for studying organelle assembly. Trends Cell Biol 3:156–161

Kamiya R (1982) Extrusion and rotation of the central-pair microtubules in detergent-treated *Chlamydomonas* flagella. Cell Motil, Supp 1:169–173

Kindle KL, Schnell RA, Fernandez E, Lefebvre PA (1989) Stable nuclear transformation of *Chlamydomonas* using a gene for nitrate reductase. J Cell Biol 109:2589–2601

Kozminski KG, Johnson KA, Forscher P, Rosenbaum JL (1992). A new motility associated with the eukaryotic flagellum. Mol Biol Cell 3:51a

Kozminski KG, Johnson KA, Forscher P, Rosenbaum JL (1993). A motility in the eukaryotic flagellum unrelated to flagellar beating. Proc Natl Acad Sci USA 90:5519–5523

Lefebvre PA, Rosenbaum JL (1986). Regulation of the synthesis and assembly of ciliary and flagellar proteins during regeneration. Ann Rev Cell Biol 2:517–546

Omoto CK, Kung C (1979) The pair of central tubules rotates during ciliary beat in *Paramecium*. Nature (Lond) 279:532–534

Omoto CK, Kung C (1980) Rotation and twist of the central-pair microtubules in the cilia of *Paramecium*. J Cell Biol 87:33–46

Omoto CK, Witman GB (1981) Functionally significant central-pair rotation in a primitive eukaryotic flagellum. Nature (Lond) 290:708–710

Purton S, Rochaix JD (1994) Complementation of a *Chlamydomonas reinhardtii* mutant using a genomic cosmid library. Plant Mol Biol 24:533–537

Remillard SP, Witman GB (1982) Synthesis, transport, and utilization of specific flagellar proteins during flagellar regeneration in *Chlamydomonas*. J Cell Biol 93:615–631

Rosenbaum JL, Moulder JE, Ringo DL (1969) Flagellar elongation and shortening in *Chlamydomonas*. The use of cycloheximide and colchicine to study the synthesis and assembly of flagellar proteins. J Cell Biol 41:600–619

Schloss JA, Silflow CD, Rosenbaum JL (1984). mRNA abundance changes during flagellar regeneration in *Chlamydomonas reinhardtii*. Mol Cell Biol 4:424–434

Silflow CD, Chisholm RL, Conner TW, Ranum LP (1985) The two alpha-tubulin genes of *Chlamydomonas reinhardtii* code for slightly different proteins. Mol Cell Biol 5:2389–2398

Silflow CD, Rosenbaum JL (1981) Multiple α- and β-tubulin genes in *Chlamydomonas* and regulation of tubulin mRNA levels after deflagellation. Cell 24:81–88

Sodeinde OA, Kindle KL (1993) Homologous recombination in the nuclear genome of *Chlamydomonas reinhardtii*. Proc Natl Acad Sci USA 90:9199–9203

Tam LW, Lefebvre PA (1993) Cloning of flagellar genes in *Chlamydomonas reinhardtii* by DNA insertional mutagenesis. Genetics 135:375–384

Williams BD, Velleca MA, Curry AM, Rosenbaum JL (1989) Molecular cloning and sequence analysis of the *Chlamydomonas* gene coding for radial spoke protein 3: flagellar mutation *pf-14* is an ochre allele. J Cell Biol 109:235–245

Youngblom J, Schloss JA, Silflow CD (1984) The two β-tubulin genes of *Chlamydomonas reinhardtii* code for identical proteins. Mol Cell Biol 4:2686–2696

A Molecular Motor for Microtubule-Dependent Organelle Transport in *Neurospora crassa*

G. STEINBERG and M. SCHLIWA[1]

1 Introduction

Kinesin and cytoplasmic dynein are two types of microtubule-dependent molecular motors that are involved in a variety of transport phenomena within eukaryotic cells. The former generally moves towards the plus end of microtubules, while the movement of the latter is minus-end-directed. Both classes of motor molecules comprise superfamilies or families of related proteins that have diversified considerably during evolution. The kinesin superfamily is defined by its founding member, the motor protein kinesin first identified in the squid *Loligo* (Vale et al. 1985). Kinesin has a globular motor domain that possesses the ATP and microtubule binding sites, an α-helical stalk, and a tail domain. Two of these "heavy chains" associate with one another via their stalk domains to form a molecule whose overall shape resembles that of a myosin molecule (Schliwa 1989). The tail domains are further associated with one copy each of a "light chain", the function of which has not been elucidated with certainty. All members of the kinesin superfamily share the putative motor head domain (usually 35–45% identity). However, the stalk/tail domains may be vastly different from each other. The general belief is that this variability within the tail domains reflects different functions or binding affinities of the respective motors. In many cases, this belief is indeed supported by experimental evidence.

2 The Kinesin Superfamily

The kinesin superfamily, of which about 30 different members are known to date, can be subdivided into four broad classes (Table 1).

The first family represents the "true" kinesins, of which the *Loligo* motor is the founding member. These molecules are thought to function primarily in organelle transport within the cell. The second family is named after the bimC gene of *Aspergillus nidulans* and comprises motors that have functions in spindle pole body or centrosome separation. Mutations in several of these genes cause defects in spindle pole separation during prophase of mitosis. The third family is named after the KAR3 gene of budding yeast. A distinctive feature of the members of this family is a reverse structural organization of the head and tail domains, i.e., the head is found at the C-terminus. Unlike other kinesins for which motility tests are available, ncd and KAR3

[1] Institute for Cell Biology, University of Munich, Schillerstr. 42, 80336 Munich, FRG.

45. Colloquium Mosbach 1994
The Cytoskeleton

Table 1. Overview of kinesin-like molecular motors

Vertebrates					
Man	KHC		clone 14	KIF3	MKLP-1
Mouse, rat	KHC				CENP-E
Xenopus		Eg5			KIF2
Echinoderms					
Sea urchin	KHC				KRP$^{85/95}$
Arthropods					
Drosophila	KHC	KLP61F	ncd	KLP64D	nod
				KLP68D	KLP98A
					KLP3A
Mollusks					
Squid	KHC				
Nematodes					
C. elegans	KHC			unc104	
Fungi					
Aspergillus		bimC	KLPA		
Yeast		cin8	KAR3		SMY1
		cut7			KIP2
		KIP1			

are minus-end-directed motors. Members of the KAR3 family may function in meiosis by converting a radial array of microtubules into a bipolar array by bundling microtubules and walking along them in the minus-end direction. The fourth family of kinesin motors is not a family in the sense that its members share certain structural or functional features; rather, the motors grouped here are characterized by the absence of related representatives in other species and may therefore be called appropriately a family of orphans. Based on mutant analyses and other data, these kinesins have a variety of functions in mitosis, meiosis, or microtubule transport (Goldstein 1993). With the exception of KRP 85/95 (Cole et al. 1993), the kinesin-like proteins have not been isolated as native molecules from cell extracts. This has only been achieved with true kinesins.

3 Organelle Transport in Fungi

One notable feature of the compilation shown in Table 1 is the apparent lack of knowledge about kinesins [and, for that matter, the mechanism(s) of organelle transport] in the kingdom of fungi. Based largely on structural and pharmacological evidence, one group of authors believes that organelle movements are microtubule-associated, while others find evidence for an involvement of actin filaments (McKerracher and Heath 1987). We have therefore undertaken a study of intracellular movement of mitochondria and other organelles in an ascomycete fungus, *Neurospora crassa*. Using cell models hitherto not employed in the study of the organelle transport machinery in fungi, it was possible to demonstrate that the vectorial movement of organelles in *N. crassa* is indeed a microtubule-dependent process (Steinberg and Schliwa 1993). By removing the cell wall and preparing flat cell fragments, *N. crassa* hyphae were transformed into a cell model with features reminiscent of animal cells, and a much clearer view of transport processes was obtained. These models made possible a high

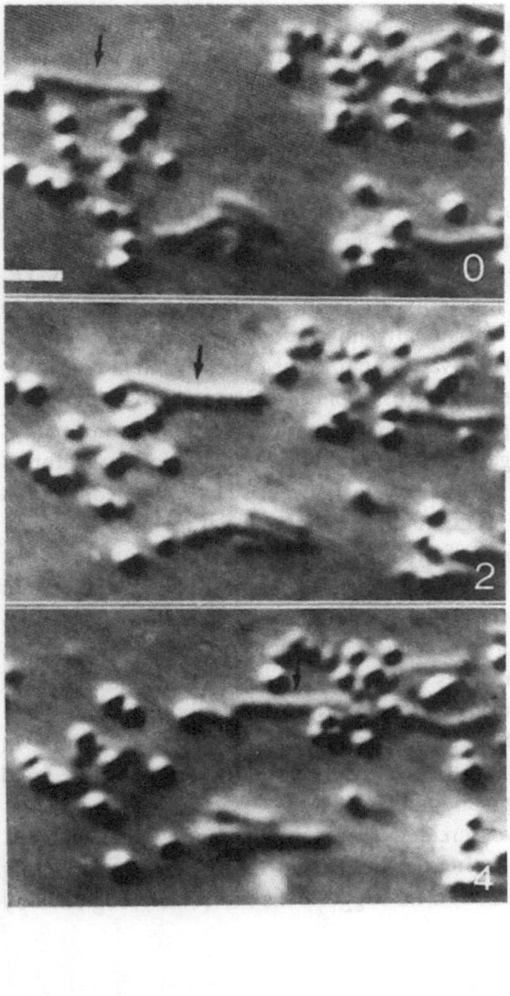

Fig. 1. Movement of mitochondria and particles in a cell wall-less mutant of *N. crassa*. Time (in s) is indicated in the *lower right*. *Arrow* denotes a moving mitochondrion. *Bar* 2.5 μm

resolution analysis by differential interference contrast optics and computer-enhanced video microscopy under a variety of experimental conditions. In all cell models the transport of organelles was vectorial and saltatory in nature (Fig. 1). The mean transport rates for mitochondria, particles, and nuclei were 1.4, 2.0, and 0.9 µm/s, respectively. Treatment with 10 µM nocodazole for 30 min caused a complete disappearance of microtubules and reversibly blocked directed transport of virtually all organelles, whereas cytochalasin D up to 20 µm was without effect. Correlative video and immunoflourescence microscopy of small fragments of wall-less mutant cells revealed a good match between microtubule distribution and the tracks of moving organelles. Thus organelle movements in *N. crassa* are microtubule-dependent and presumably require the presence and force-generating capabilities of a microtubule-associated motor molecule.

4 Identification of a *N. crassa* Molecular Motor

To begin an analysis of the molecular basis of organelle transport in *N. crassa*, we have prepared cytoplasmic extracts of 12–16 h *N. crassa* hyphae and have partially purified an ATP and microtubule-dependent motor activity. This motor activity binds to microtubules in the absence of ATP or in the presence of excess AMPPNP over ATP. When released from microtubules with ATP this motor activity promotes microtubule gliding over coverslips coated with motor molecules at a rate of 2–2.5 µm/s. Using flagellar axonemes of *Chlamydomonas*, the polarity of this motor activity was determined to be directed towards the microtubule plus-end. Microtubule gliding is optimal in the presence of ATP, but can also be induced by GTP, CTP, and UTP. Vanadate and NEM markedly affect the ability of microtubules to bind to motor-coated coverslips, but occasionally microtubules are seen to move at high rates even in the presence of 100 µM vanadate or 5 mM NEM. The motor activity co-fractionates with a polypeptide doublet of 105/108 kDa in sucrose gradients. Blot-purified polyclonal antibodies against either polypeptide of the doublet recognize both bands in Western blots, suggesting that these two polypeptides are isoforms of the same protein. The nature of the difference between these two isoforms is presently under investigation. Further fractionation of the material released from microtubules with ATP on a superose-6 gel filtration column yields motor-active fractions of the 105/108 kDa doubled at over 80% purity. No associated "light chains" are found. Based on a combination of sucrose density gradient fractionation and gel filtration, the most likely composition of the native motor molecule is two subunits of the 105/108 kDa polypeptide. The properties of this motor enzyme are summarized in Table 2. Immunofluorescence microscopy with monospecific antibodies against the 105/108 kDa protein show staining of small vesicles associated with microtubules in cell fragments of wall-less *N. crassa* cells. Thus the 105/108 kDa protein is a good candidate for the motor that drives organelle movements in *N. crassa*.

Using the polyclonal antibody, we screened a λgt 11 expression library of *N. crassa* and obtained a partial clone spanning the C-terminal ~80% of an open reading frame. The sequence information obtained so far suggests the molecule recognized by

Table 2. Comparison of native isolated kinesins and the *N. crassa* motor

		N. crassa motor	Kinesin
V MT-gliding	(µm/s)	2.5–3	~0.5
V bead-movement	(µm/s)	1.5	~0.5
MgATP	(%)	100	100
MgGTP	(%)	50	80–100
MgCTP	(%)	10–20	10–30
MgUTP	(%)	~10	10–40
MgITP	(%)	0	30–50
Direction		Plus	Plus
Vanadate-inhibition	(µm)	>70	>50
NEM-inhibition	(mM)	>2–5	>2–5
Vanadate/UV-cleavage		NO	NO
Calculated MW	(kDa)	~230	~380
Composition	(kDa)	2 x~105	2 x~110–130 2 x~50–80
Sedimentation constant Sw20	(S)	8.8	9.5
Specific ATPase activity – microtubules	[µmol/min/mg]	0.006–0.0,021	Mean ~0.018
Specific ATPase activity + microtubules	[µmol/min/mg]	0.032–0.118	Mean ~0.1
MT activation		5.1–5.6 x	Mean ~5.6

The data for kinesins were compiled from sources referenced in Goldstein (1993).

this antibody as possessing a N-terminal head closely related in sequence to other kinesins. However, the rest of the molecule bears no sequence similarity to any other kinesin or kinesin-like protein.

The information summarized here shows the molecular motor that most likely drives organelle transport in *N. crassa* to be kinesin-like, but with some unique properties that sets it apart from all other kinesins described so far. The rate of transport is five to six times higher than that of other "true" kinesins; only the motor protein identified in *Dictyostelium* possesses similarly high motility rates (McCaffrey and Vale 1989), but is immunologically distinct from the *N. crassa* motor. The physicochemical characteristics of the *N. crassa* motor suggest it to be a much more compact molecule than other kinesins. Finally, its sequence outside the head region is unique among all kinesins and kinesin-like proteins identified so far. This is remarkable, because in terms of its function this motor performs the tasks believed to be carried out by the "true" kinesins, all of which exhibit related stalk and tail domains. The *N. crassa* motor may therefore represent a new class or family of kinesin motors that may be representative of fungi.

Acknowledgments. We thank Dr. W. Neupert and M. Braun for allowing us to use their *N. crassa* culture facility, Dr. H. Weiss for a gift of the cDNA library, S. Seiler for active help with some of the experiments, and Drs. U. Euteneuer, E. Granderath, and H. Siess for helpful discussions and advice throughout the course of this work. Supported by the Deutsche Forschungsgemeinschaft.

References

Cole DG, Chinn SW, Wedaman KP, Hall K, Vuong T & Scholey JM (1993) Novel heterotrimeric kinesin-related protein purified from sea urchin eggs. Nature 366:268–270

Goldstein LSB (1993) With apologies to Sheherazade: tails of 1001 kinesin motors. Annu Rev Genet 27:319–351

McCaffrey G & Vale RD (1989) Identification of a kinesin-like microtubule-based motor protein in *Dictyostelium discoideum*. EMBO J 8:3229–3234

McKerracher LJ & Heath IB (1987) Cytoplasmic migration and intracellular organelle movements during tip growth of fungal hyphae. Exp Mycol 11:79–100

Schliwa M (1989) Head and tail. Cell 56:719–720

Steinberg G & Schliwa M (1993) Organelle movements in the wild-type and wall-less fz; sg; os-1 mutants of *N. crassa* are mediated by cytoplasmic microtubules. J Cell Sci 106:555–564

Vale RD, Reese TS & Sheetz MP (1985) Identification of a novel force-generating protein, kinesin, involved in microtubule-based motility. Cell 42:39–50

The Actomyosin Interaction

K. C. HOLMES[1]

1 Introduction

An atomic model for the rigor complex of F-actin and the myosin head has been obtained by combining the molecular structures of the individual proteins (Rayment et al. 1993a) (Holmes et al. 1990) (Lorenz et al. 1993) with the low resolution electron density maps of the actomyosin complex derived by cryoelectron microscopy (Rayment et al. 1993b) (Schröder et al. 1993). A model for the actomyosin interaction has been proposed in which the actin binding sites and nucleotide binding sites of S1 are functionally linked by a cleft in the 50K domain of S1 which is thought to close on binding to F-actin (Rayment et al. 1993b). The closing of the cleft is likely to be an essential part of the weak/strong sequence of the actomyosin interaction. The initial actomyosin interaction probably involves only part of the surface, the full rigor complex develops on closure of the cleft. The obligatory sequence *stereo specific-weak/strong* prevents the cross-bridge from binding unstrained in the strong state. Tropomyosin in the "off" state appears to inhibit the closing of the cleft.

2 The Cross-Bridge Cycle

Muscle contraction occurs when two sets of interdigitating filaments, the thin actin filaments and the thick myosin filaments slide past each other. A widely accepted theory to explain this sliding process is the cross-bridge theory of muscle contraction (see Huxley 1969; 1974). This theory suggests that the sliding process is driven by the myosin cross-bridges which extend from the myosin filament and cyclically interact with the actin filament by a rowing motion as ATP is hydrolyzed. Based on the crystal structure of globular actin (Kabsch et al. 1990) the refined structure of the actin filament is now available (Lorenz et al. 1993). The cross-bridges comprise a part of the myosin molecule, namely subfragment-1 of heavy meromyosin (S1), the structure of which has been solved (Rayment et al. 1993a). This study shows the S1 to be tadpole-like in form, with an elongated head, containing a 7-stranded β sheet and numerous associated α-helices, which has the actin binding sites and nucleotide binding sites in opposite sides and is connected to an extended α-helix which forms the C-terminal tail and binds the two light chains.

Evidence coupling a rowing-like motion of the cross-bridge (or part of the cross-bridge) to force generation was prevented by Huxley et al. (1981) and Irving et al. (1992). However, structural work has met difficulties in substantiating the expected

[1] Max Planck Institute for Medical Research, 69120 Heidelberg, FRG.

45. Colloquium Mosbach 1994
The Cytoskeleton

large-scale rowing-like motion of the S1 which led to the proposal (Goody et al. 1983; Cooke 1986) that the site of the rowing-like motion which produces movement (the motor unit) was in the distal part of the S1 molecule and not very close to actin. Furthermore, other structural studies showed that the ATP binding site was remote from the actin binding site (Botts et al. 1989; Sutoh et al. 1989). One ends up with a scheme in which the three essential functionalities of the cross-bridge, actin binding, nucleotide binding, and the motor region are spacially disjoint but functionally linked (Wray et al. 1988). The molecular model of the actomyosin complex (Rayment et al. 1993b) substantiated these ideas and immediately suggested a physical basis for the proposed rowing-like motion. Moreover, Rayment et al. were able to suggest the basis of the linkage between the nucleotide binding site and the actin binding site.

S1 binds tightly to actin filaments but also binds and hydrolyzes ATP. ATP binding brings about a rapid dissociation of S1 from actin. Thus S1 can bind either to actin or ATP but to both only transiently. The presence of the ATP γ-phospate is crucial for dissociation. ADP alone has little effect. Solution kinetic observations were very important in establishing the relationship between the hydrolysis of ATP and the generation of force. A key feature of this process is the observation that transduction of the chemical energy released by the hydrolysis of ATP into directed mechanical force occurred during product release (ADP and inorganic phosphate, P_i) rather than during the hydrolysis step itself (Lymn and Taylor 1971). Without actin, myosin is product-inhibited. Mg-ATP rapidly dissociates the actomyosin complex by binding to the ATPase site of myosin; free myosin then hydrolyzes ATP and forms a stable myosin-products complex; actin recombines with this complex and dissociates the products, thereby forming the original actin-myosin complex. Force is generated during the last step.

This cycle has been elaborated by Eisenberg and Green (1980) to incorporate the fact that release from actin is not an obligatory step in the hydrolysis of ATP. In their cross-bridge model, each sweep of an oar-like cross-bridge cycle is not linked rigorously to a biochemical event, but the myosin cross-bridge alternates between a weak binding conformation and a strong binding conformation, which differ in the structural way in which they interact with actin. In the muscle lattice, the strongly bound state is created in a strained conformation from the weakly bound form. Since the strongly bound state has *at equilibrium* a different gross orientation to the actin filament (45°) but is tethered by its attachment to the myosin filament to remain in the initial 90° position, it is under strain. Release of this strain brings about the rowing-like sweep. Hill (1974) emphasized that this force-producing event, often referred to as the power stroke, should not occur *between* two states but *within* a state so that the stored energy can be smoothly given up into the muscle filament lattice. Alternatively, one pictures the power stroke as a series of closely spaced states with no activation energies between them.

The Eisenberg and Green cycle has provided the basis for most current thinking about contraction mechanisms. Central to this kind of cycle is the idea that the cross-bridge must first bind in an unstrained conformation (weak) and then undergo an isomerization to the strained (strong) form. The process is modulated by the status of the nucleotide binding site, in particular by the presence or absence of the γ-phosphate.

This is the heart of the contraction process. A key question which we wish to address is how the transition from weak to strong binding comes about.

3 Why You Need Weak to Strong

The initial attachment is weak in the sense that the bound cross-bridge is in rapid equilibrium with the dissociated form. However, it should be *stereo specific binding*, otherwise it would not be possible to provide a structural basis for the ensuing isomerization to the strong binding state. Furthermore, the stereo specific weak binding to acting is required to *enable* (catalyze) the transition to the strong binding state. In the absence of actin the strong binding conformation of S1 is simply kinetically not available. This sequence provides the necessary vectoriality to the process (Jencks 1980). Without this property the cross-bridge could and would bind unstrained in the strong state and nothing would happen. This scheme is not a unique paradigm for achieving a strained bound cross-bridge. However, the two-step binding of S1 to actin is in accordance with the biochemic data (Taylor 1991; Geeves 1992) and provides a starting model for our discussion. Geeves's A and R states correspond to the stereo specific weak binding and strong binding states discussed below.

4 The Low Ionic Strength Interaction Is a Pre-Weak State

While the weak state we discuss here is conceptually similar to that discussed by Eisenberg and Green, it may have little to do with the weak states found at low ionic strength (Brenner et al. 1982) since, for example, it has been shown by Pollard et al. (1993) that at low ionic strength the cross-bridges take up all possible orientations with respect to the actin filament and are therefore not specifically bound to actin. The interactions are electrostatic in nature and will not, in general, be competent to make the transition to the strong state because their orientation is not right. Such non-competent electrostatic *weak states* (Wray et al. 1988) should be viewed rather as the cross-bridge making short-lived collision complexes with actin.

5 Summary of the Results of Rayment et al. 1993b

The atomic model arrived at by combining the structures of actin and myosin with the cryoelectron microscope reconstruction of "decorated actin" is shown in Fig. 1. In the following reference will be made to the chicken sequence of myosin S1 (Maita et al. 1991).

Myosin S1 can be split by trypsin into three fragments (Mornet et al. 1981) (25, 50, 20K from N to C-terminus) which were thought to be domains. Rayment et al. (1993a) show that these "domains" do not reflect structural divisions, but rather that

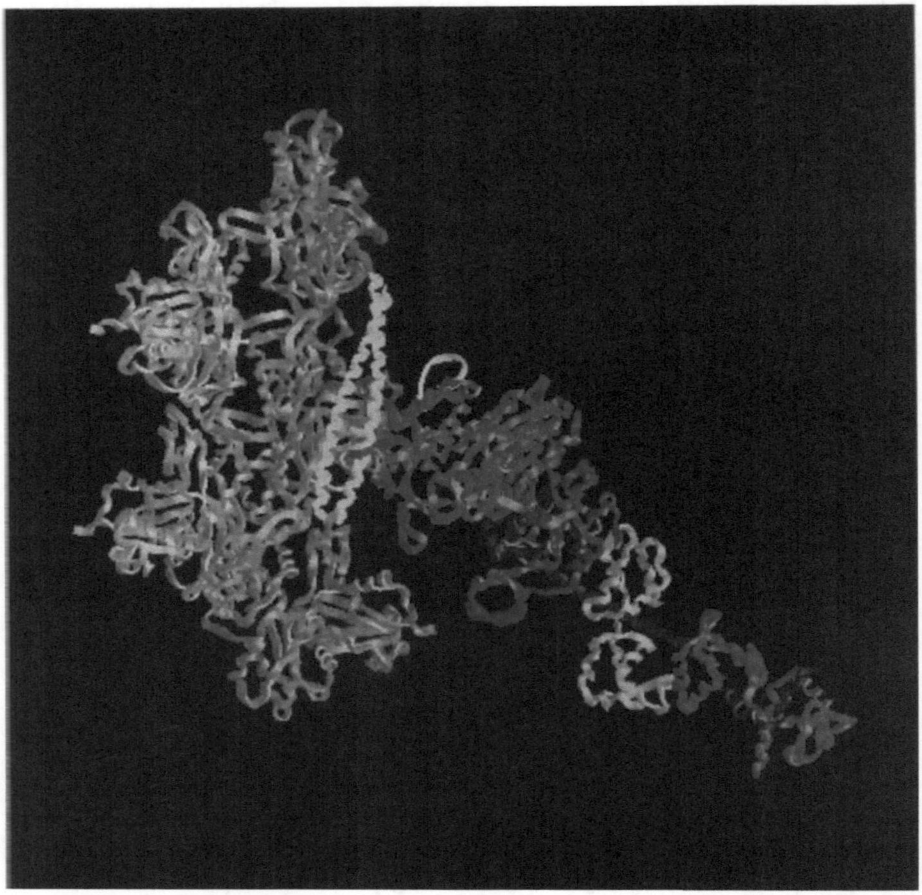

Fig. 1. Atomic model of the actomyosin complex (Rayment et al. 1993b) in a ribbon representation. Shown are five actin molecules and one myosin S1. The actin helix is *in the plane* of the diagram. The barbed end (Z-line) is *at the bottom*. Note the two light chains which enfold the long α-helical C-terminal tail of myosin S1. Also shown is a tropomyosin molecule in the "rigor" position. (Figure prepared with GRASP: A. Nicholls)

the junctions between putative domains arise from loops which are not seen (disordered) in the crystal structure and are available for proteolysis. However, the terminology (50K etc.) is now well established and will be used in what follows. Rayment et al. show that the 50K fragment is further split into two by a long cleft. The two subdomains are referred to as the upper and lower 50K fragments (or domains). The upper 50K fragment contains a 7-stranded β sheet and forms part of the nucleotide binding site. The lower 50K fragment contains the primary actin binding site. (Fig. 2).

The full actomyosin interaction is probably built from components of both the upper and lower 50K fragments. These interactions are found on both sides of the cleft.

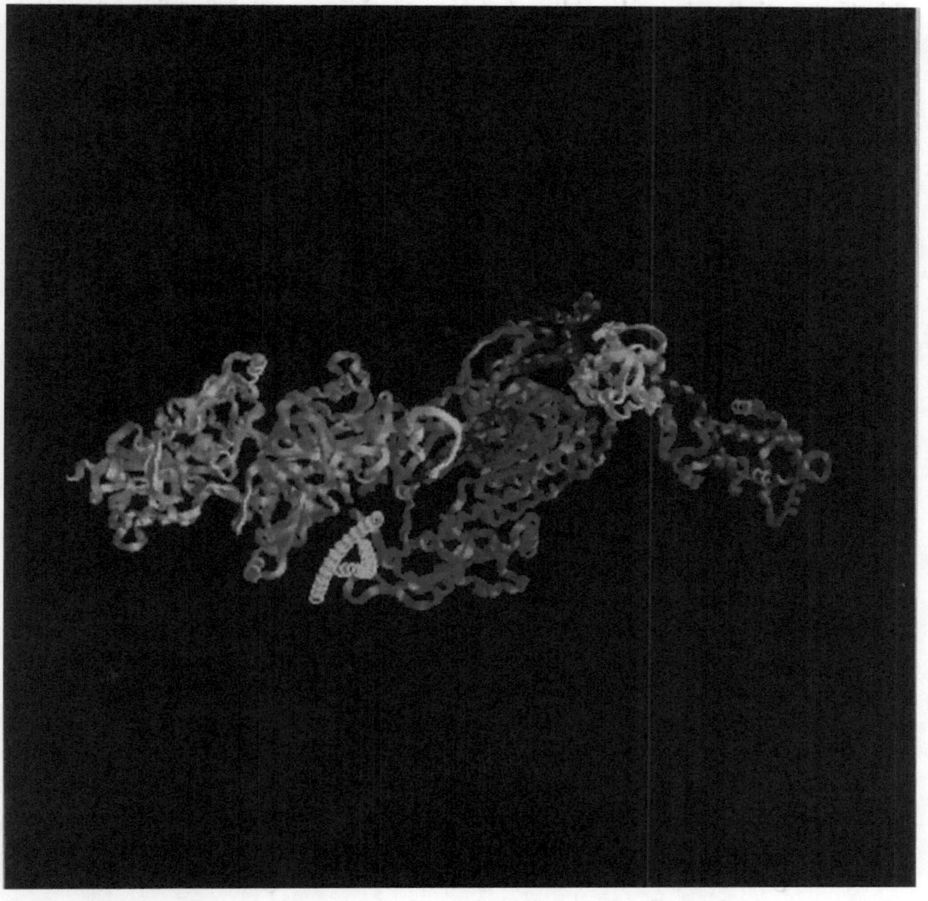

Fig. 2. A view of the actomyosin interface approximately at right angles to the view in Fig. 1 seen from above. Note the cleft between the 50K upper and lower domains. The tropomyosin is in contact with the 50K upper domain. (Figure prepared with GRASP: A. Nicholls)

Examination of this interface suggests that there are potentially three types of interaction with actin: initially an ionic interaction involving the flexible 50K–20K junction; secondly, a stereo specific interaction involving hydrophobic residues which we associate with the weak interaction; thirdly a strengthening of this interaction by the recruitment of more hydrophobic interactions allowed by the coming together of the 50K upper and lower fragments to produce the strong interaction.

The segment between residues 626 and 647 (50K/20 K junction) is disordered in the three-dimensional structure and contains five lysines and nine glycines. However, these lysine residues are protected from proteolysis in the presence of actin which suggests that while they are flexible in solution they are physically protected by actin (Mornet et al. 1979) or only adopt a distinct conformation when bound in the actomyosin interface. From their location the residues between 626 and 647 may interact with the negatively charged residues 1–4 on actin. It is expected that this interaction will be predominately ionic (five lysines in the loop and four carboxylic acid groups clustered at the N-terminus of actin, and therefore would be sensitive to ionic strength. It is possible that this component of the structure is partially responsible for the ionic strength-dependent "weak binding" established as a characteristic of low ionic strength actomyosin interaction by Brenner et al. (1982). This is the line with many observations whereby the initial weak interaction is seen to orient the myosin head so that its nose is close to actin without there being any specific geometrical relationship between actin and myosin.

The main contact of myosin with actin involves the heavy chain segment from 525–552. This consists of a helix, that extends from 516 to 542, a loop from 543–546 and a second helix from 547–558. The first helix, of which only part is in the actin binding site, contains a prominent bulge at residue 529 which fits into a cavity on the surface of actin. These two helices run at an angle of 10° to each other and are located at the end of the lower 50K domain. These secondary structural elements are in close proximity to residues 341 and 354 and 144 to 146 of actin. These are the main components of the actomyosin interaction which can be discerned with the cleft in the "open" configuration (as it is found in the crystal structure).

In addition to the loop at the 50K/20K junction, a loop between 405 and 415 extends towards the actin filament. If the cleft between the upper and lower 50K fragments should close, this loop would be well placed to interact with actin, probably around the pair of exposed proline residues (332–333) on actin subdomain 3. The importance of this loop in the function of the protein has been implicated from genetic studies of familial hypertrophic cardiomyopathy that show that mutation of residue Arg 405 to Gln in cardiac myosin (based on the chicken sequence) is one of the causes of this disease state (Geisterfer-Lowrance et al. 1990). This is consistent with the loop being an essential part of the strong interaction. The cleft between the upper and lower 50K fragments is also apparently the line of communication between the actin binding site and nucleotide binding site, since a closing of this cleft would have consequences in the nucleotide binding site.

6 A Mechanism for Controlling the Weak/Strong Transition

The structure of myosin at the interface suggests a mechanism for the transition from weak to strong binding. As noted, the stereo specific weak binding probably comprises the lower 50K fragment (i.e., myosin residues 525–552 and 547–558 together with ionic interactions between the 626–647 loop and the actin N-terminus and 99–100). However, immediately adjoining the 627–646 loop the polypeptide chain makes an excursion across the cleft separating the upper and lower fragments to build a small a-helical plug (600–625) which is attached to the 50K upper fragment. It seems likely that the binding of the loop to the surface of actin, and possibly its subsequent naturation, would tend to close the cleft by shortening the loop. The fact that this binding interaction is coupled to loop naturation means that it is entropically disfavored and perhaps could only take place after the first stereo specific weak interac-

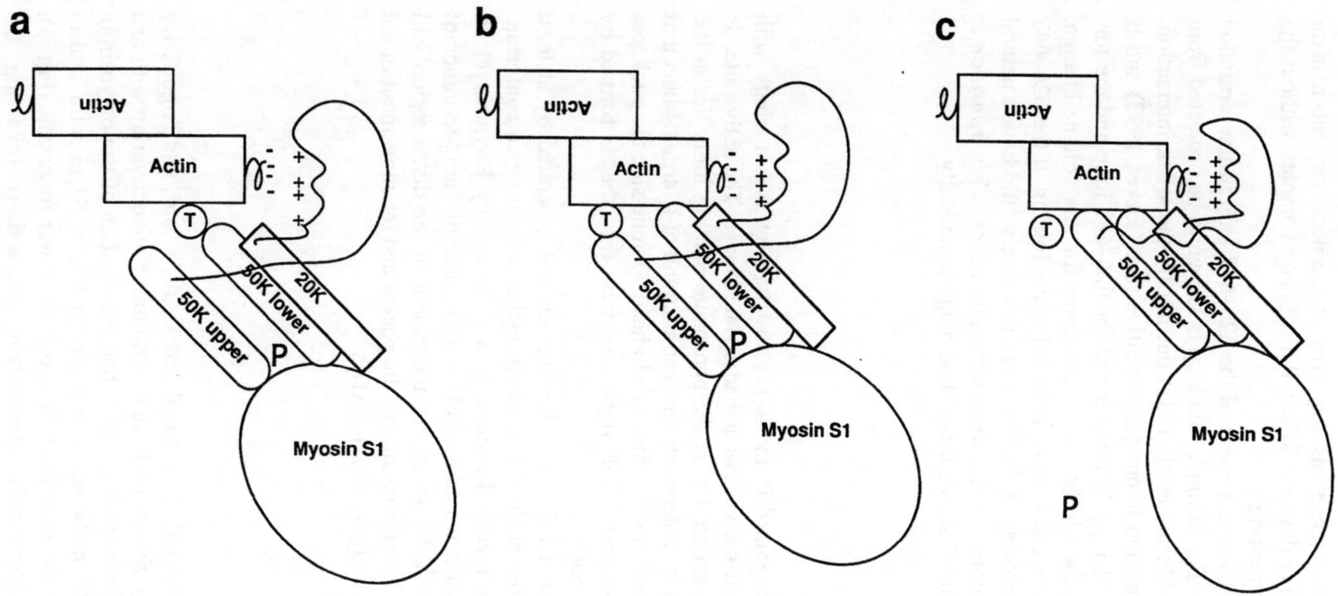

Fig. 3 a–c. A diagram looking down the z-axis of the thin filament towards the z-line showing the relative dispositions of actin, tropomyosin, and S1 (Rayment et al. 1993b, Poole et al. 1994; Lorenz et al. 1994). The N-terminus of actin is shown as a short helix. **a** The "off" state – the tropomyosin sits in the cleft between the 50K upper and lower domains. Note that electrostatic interactions between the 50K–20K loop and the N-terminus of actin would not be influenced by tropomyosin. **b** "On" state – note steric hindrance between the 50K upper fragment and tropomyosin. **c** Rigor complex (the closing of the cleft is hypothetical) – the tropomyosin lies outside but adjacent to the myosin binding site. A possible mechanism for controlling the hypothesised closing of the cleft would be an interaction between the 50K–20K loop and the N-terminus of actin brought about by the myosin binding to actin. The interaction between the loop and the N-terminus may shorten the loop and favor closing of the cleft

tion. Therefore, the loop and its interactions may provide the necessary mechanism for ensuring that the rigor binding does not follow the collision complex without the stereo specific weak binding first taking place.

There is good evidence for the involvement of the 626–647 loop in the formation of the strong (rigor) interaction. For example, the loop is known to be protected from hydrolysis by this interaction (Mornet et al. 1981) but not by the weak interaction. Moreover, the weak-strong transition is pressure-sensitive (see Geeves 1992) and is associated with a large volume change. The folding of the 626–647 loop and the formation of four to five ion pairs would be ready candidates for this volume change. Furthermore, if formation of the weak binding state led to the folding of the 626–647 loop, and this in turn led to the closing of the 50K cleft, then one would have a natural explanation of the sequential control of the weak-strong transition. *This transition is the quintessence of muscle.* Figure 3 shows these ideas diagrammatically.

7 The ATP Binding Pocket

The ATP binding site was recognized in the X-ray structure by its homology with other nucleotide binding proteins such as ras and adenylate kinase. The active site, as identified by the location of a sulfate ion in the phosphate binding loop, lies at the bottom of a pocket. The pocket is centered on the central strand of a large β-sheet that forms the major tertiary structural motif of the myosin head. On one side the pocket is formed by segments of the N-terminal 25K region, whereas the other is formed by segments from the central 50K region.

The nucleotide binding site and the actin binding site are separated by at least 35 Å. In the crystal structure, the nucleotide binding pocket is in an open conformation and faces away from the F-actin filament. It is estimated by Rayment et al. (1993a) that closure of this pocket would result in a movement of the C-terminus of the heavy chain by at least 50 Å. However, the movement of the distal region will also depend on actin binding, since the closing of the pocket and the movement of the distal tail cannot be achieved simply by binding ADP.

8 Control by Tropomyosin

The thin filament has three identifiable structural states: "off", "on", and "rigor". By analysis of the low-angle X-ray diffraction from rabbit muscle at nonoverlap with and without Ca^{2+}, models can be derived which show that the position of tropomyosin in the "off" state (no Ca^{2+}) would inhibit stereo specific binding of S1 to actin, and in particular would sit in the cleft between the 50K upper and lower fragments (Fig. 3). It would, however, allow nonspecific electrostatic interactions involving the 50K–20K loop and the N-terminus of actin. In the presence of Ca^{2+} a stereo specific weak interaction would be permitted but the closure of the cleft would lead to some steric clashes with tropomyosin (Lorenz et al. 1994; Poole et al. 1994). A further small movement of the tropomyosin would permit the closure of the cleft and the full rigor interaction. It is attractive to equate these three structural states with the states "blocked", "closed", and "open" identified by McKillop et al. 1993.

9 The Power Stroke

It seems probable that the power stroke involves a rowing-like motion of the distal C-terminal tail carrying the two light chains, and that this motion is produced by the motor region which comprises part of the 20K region and parts of the 25K region lying immediately distal to the nucleotide binding pocket and perhaps the nucleotide pocket itself.

A major contribution to the power stroke may arise from changes in the myosin heavy chain induced by nucleotide binding and release, a mechanism for which arises naturally out of the structure of S1 (Rayment et al. 1993a). However, the cross-bridge is capable of producing force *even before* phosphate release (Cooke et al. 1985); see Brenner (1990) for review. Moreover, there is evidence for conformational changes in the region of the molecule distal to the nucleotide binding site, induced just by actin binding (Bertrand et al. 1992). Therefore the induction of the power stroke appears to have contributions from both phosphate release and actin binding. Wray et al. (1988) envisaged a synergism between actin binding and phosphate release as being responsible for the straining of the motor region (and hence making the tail turn). The structure presented by Rayment et al. (1993a) makes this idea quite credible, since the closing of the 50K cleft can be communicated directly to the helix 680–710 which contains the reactive SH1 SH2 groups and which is probably near the focus of the motor region. One knows from chemical studies (Huston et al. 1988) that the structure of this part of the molecule must be quite different on binding nucleotide, moreover, it is close to the nucleotide pocket. Furthermore, the tail of the molecule lies immediately distal to this region. It is, therefore, in a good position to respond to both the status of the actin binding region and the nucleotide binding region and to respond by regulating the sweep angle of the tail.

The atomic model presented by Rayment et al. (1993a) offers immediate implications for the changes that must occur during the active parts of the cycle. In particular it suggests that the segment of myosin that forms the tight interaction with actin does so in only one orientation, and hence that the motor domain or segment of the structure responsible for the power stroke lies some distance from this interface. Initially, the myosin forms a stereo specific *weak* interaction with actin involving the lower 50K fragment. This interaction still has a high off rate. Sometimes, however, it makes the transition to strong binding by forming a structured interaction between the 20K-50K loop and the actin N-terminus which in turn allows the gap between the upper and lower 50K fragments to close to produce *strong* binding. This interaction strains both the *motor region* and the *nucleotide binding site*. The strain results in a raising of the off rate for the γ phosphate, which is released. An untethered myosin molecule would now go about the conformational changes associated with opening the nucleotide binding pocket and releasing the ADP. At this stage, loss of ADP would terminate the power stroke returning the molecule in its rigor state. However, myosin is tethered by the S2 and can only undergo this conformational change at a rate determined by the actomyosin lattice movements.

Lombardi et al. (1992) have shown that under certain conditions cross-bridges may release during the power stroke and reattach to actin to generate force without rebinding ATP. Such cross-bridges have lost their phosphate but not their ADP. If strong actin binding can alone produce an alteration of the angle of the tail, a structural model with *controlled sequential binding* to actin provides a ready explanation of this phenomena. The result of Sleep and Hutton (1980) showing that there is an activated ADP state in the cycle not available just by incubating actomyosin with ADP is also consistent with this hypothesis.

Acknowledgments. I gratefully acknowledge the help of Ivan Rayment and Hazel Holden in allowing me the use of the myosin S1 coordinates and for many helpful discussions.

References

Bertrand R, Derancourt J, Kassab R (1992) Molecular movements in the actomyosin complex: F-actin-promoted internal cross-linking of the 25- and 20-kilodalton heavy chain fragments of skeletal myosin-subfragment-1. Biochemistry 31:12219–12226

Botts J, Thomason JF, Morales MF (1989) On the origin and transmission of force in actomyosin subfragment 1. Proc Natl Acad Sci USA 86:2204–2208

Brenner B (1990) Muscle mechanics and biochemical kinetics. Topics in molecular and structural biology (ed Squire JM) 13:77–143

Brenner B, Schoenberg M, Chalovich JM, Greene LE, Eisenberg E (1982) Evidence for cross-bridge attachment in released muscle at low ionic strength. Proc Natl Acad Sci USA 79:7288–7291

Cooke R (1986) The mechanism of muscle contraction. CRC Crit Rev Biochem 21:53–118

Cooke R, Pate E (1985) The effects of ADP and phosphate on the contraction of muscle fibres. Biophys J 48:789–798

Eisenberg E, Green LE (1980) The relation of muscle biochemistry to muscle physiology. Ann Rev Physiol 42:293–309

Geeves MA (1992) The actomyosin ATPase: a two-state system. Phil Trans R Soc Lond B 336:63–70

Goody RS, Holmes KC (1983) Cross-bridges and the mechanism of muscle muscle contraction. Biochim Biophys Acta 726:13–39

Hill TL (1974) Theoretical formulation for the sliding filament theory of contraction in striated muscle. Part 1. Prog Biophys Molec Biol 28:267–340

Huston EE, Grammer JC, Yount RG (1988) Flexibility of the myosin heavy chain: direct evidence that the region containing SH1 and SH2 can move 10 Å under the influence of nucleotide binding. Biochemistry 27:8945–8952

Huxley AF (1974) Muscular contraction. J Gen Physiol (Lond.) 243:1–43

Huxley HE (1969) The mechanism of muscular contraction. Science 164:1356–1366

Huxley HE, Simmons RM, Faruqui AR, Kress M, Bordas J, Koch MHJ (1981) Millisecond time-resolved changes in X-ray reflections from contracting muscle during rapid mechanical transients, recorded using synchrotron radiation. Proc Natl Acad Sci USA 78:2297–2301

Irving M, Lombardi V, Piazzesi G, Ferenczi MA (1992) Myosin head movements are synchronous with the elementary force-generating process in muscle. Nature (Lond) 357:156–158

Jencks WP (1980) The utilisation of binding energy in coupled vectorial processes. Advances in enzymology and related areas of molecular biology 51:75–106

Kabsch W, Mannherz HG, Suck D, Pai EF, Holmes KC (1990) Atomic structure of the actin: DNase I complex. Nature 347:37–44

Lombardi V, Piazzesi G, Linari M (1992) Rapid regeneration of the actin-myosin power stroke in contracting muscle. Nature (Lond) 35:638–641

Lorenz M, Popp D, Holmes KC (1993) Refinement of the F-actin model against X-ray fiber diffraction data by the use of a directed mutation algorithm. J Mol Biol 234:826–836

Lorenz M, Popp D, Poole KV, Rosenbaum G, Holmes KC (1994) An atomic model of the unregulated thin filament obtained by X-ray diffraction from orientated actin-tropomyosin gels. Molecular motors. Structure, mechanics and energy transduction. Airlie, Virginia, Biophysical Society.

Lymn RW, Taylor EW (1971) Mechanism of adenosine triphosphate hydrolysis of actomyosin. Biochemistry 10:4617–4624

Maita T, Yajima E, Nagata S, Miyanishi T, Nakayama S, Matsuda G (1991) The primary structure of skeletal muscle heavy chain: IV sequence of the rod, and the complete 1938-residue sequence of the heavy chain. J Biochem (Japan) 110:75–87

McKillop DFA, Geeves MA (1993) Regulation of the interaction between actin and myosin subfragment 1: evidence for three states of the thin filament. Biophys J 65:693–701

Mornet D, Bertrand R, Pantel P, Audemard E, Kassab R (1981) Structure of the acto-myosin interface. Nature (Lond) 292:301–306

Mornet D, Pantel P, Audemard E, Kassab R (1979) The limited tryptic cleavage of the chymotriptic S1: an approach to the characterisation of the actin site in myosin heads. Biochem Biophys Res Comm 89:925–932

Pollard TD, Bhandari D, Maupin P, Wachstock D, Weeds A, Zol H (1993) Direct visualization by electron microscopy of the weakly-bound intermediates in the actomyosin ATPase cycle. Biophys J 64:454–471

Poole KV, Holmes KC, Rayment I, Lorenz M (1994) Control of the actomyosin interaction. Molecular motors. Structure, mechanics and energy transduction, Airlie, Virginia, Biophysical Society

Rayment I, Rypniewski WR, Schmidt-Base K, Smith R, Tomchick DR, Benning MM, Winkelmann DA, Wesenberg G, Holden HM (1933a) The three-dimensional structure of a molecular motor, myosin subfragment-1. Science 261:50–58

Rayment I, Holden HM, Whittaker M, Yohn CB, Lorenz M, Holmes KC, Milligan RA (1993b) Structure of the actomyosin complex and its implications for muscle contraction. Science 261:58–65

Schröder RR, Manstein DJ, Jahn W, Holden HM, Rayment I, Holmes KC, Spudich JA (1993) The interaction of the *Dictyostelium* myosin head with actin: a structural model of decorated actin. Nature 364:171–174

Sleep JA, Hutton RL (1980) Exchange between inorganic phosphate and adenosine 5'-triphosphate in the medium by actomyosin subfragment 1. Biochemistry 19:1276–1283

Sutoh K, Tokunaga M, Wakabayashi T (1989) Electron microscope mappings of myosin head with site-directed antibodies. J Mol Biol 206:357–363

Taylor EW (1991) Kinetic studies on the association and dissociation of myosin subfragment 1 and actin. J Biol Chem 266:294–302

Wray J, Goody RS, Holmes KC (1988) Towards a molecular mechanism for the cross bridge cycle. Advances in experimental medicine and biology 226:49–59

Structure and Functions of Profilin, Insights at Atomic Resolution

T. D. POLLARD[1]

Profilins are small, highly abundant cytoplasmic proteins that bind actin, poly-L-proline, and polyphosphoinositides, raising the possibility that they are second messengers to carry information between the phosphoinositide signaling pathway and the cytoskeleton. Profilin is required for a normal actin cytoskeleton in both *Drosophila* (Cooley et al. 1992) and yeast (Haarer et al. 1990). Originally, profilins were considered to be inhibitors of actin polymerization (Carlsson et al. 1977), but now they appear to promote assembly in at least two different ways. In addition to their roles in the cell, the plant profilins are now recognized as major human allergens (Valenta et al. 1991). Many species, including humans (Honore et al. 1993), have more than one profilin isoform (Kaiser et al. 1986) with different properties (Machesky et al. 1990) that are expressed from different genes (Binette et al. 1990; Pollard and Rimm 1991; Haugwitz et al. 1991).

We have reviewed the development of ideas about profilin (Machesky and Pollard 1993) and the evolution of the profilin family (Pollard and Quirk 1994). Here I will summarize insights provided by the recent publications of the atomic structures of several profilins.

Our group has determined the atomic structure of *Acanthamoeba* profilin-I by multidimensional NMR (Vinson et al. 1993), amoeba profilin-I and profilin-II by X-ray crystallography (Fedorov et al. 1994a), and human profilin by X-ray crystallography (Fedorov et al. 1994b). Metzler et al. (1993) determined the solution structure of human profilin by NMR and Schutt et al. (1993) determined the crystal structure of bovine profilin bound to beta-actin. All of the profilins have the same fold (Fig. 1). They consist of a 6-stranded β-sheet flanked on one side by N- and C-terminal α-helices and on the other side by a two helices and a 2 stranded β-sheet. The vertebrate profilins have 14 more residues than the protozoan and yeast profilins. These extra residues are accommodated in surface loops between elements of secondary structure. These loops are the most flexible parts of the vertebrate profilins (Constantine et al. 1994). The conservation of structure at atomic resolution is remarkable, given the minimal conservation of primary structure. In pairwise comparisons, profilins from disparate species share about 25% of their amino acids. However, among the known profilins, only one residue, tryptophan-2 (amoeba numbering) is conserved across the phylogenetic tree among animals, protozoa, fungi, plants, and vaccinia virus (see Pollard and Quirk 1994).

[1] Department of Cell Biology and Anatomy, Johns Hopkins Medical School, 725 N. Wolfe Street, Baltimore, MD 21205, USA.

45. Colloquium Mosbach 1994
The Cytoskeleton

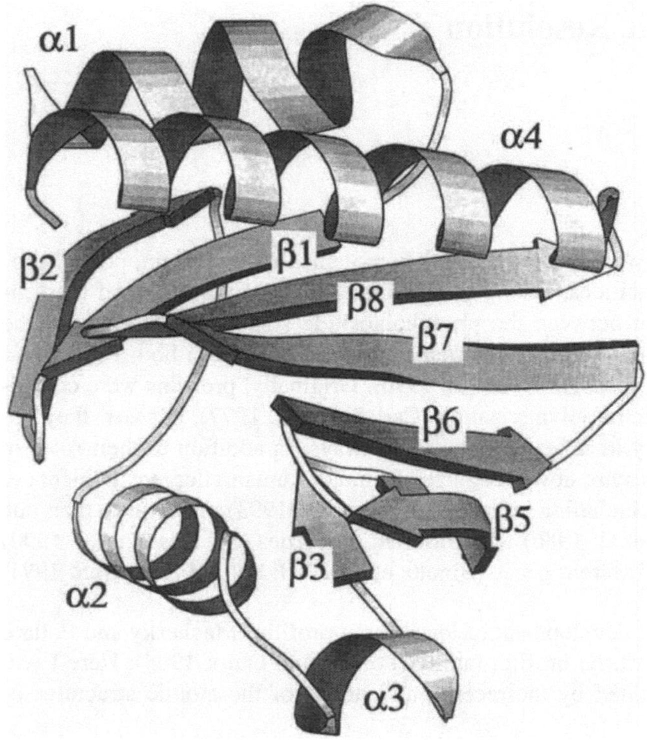

Fig. 1. Ribbon model of the secondary structure of *Acanthamoeba* profilin based on the X-ray coordinates of Fedorov et al. (1994a). Poly-L-proline binds to residues between α1 and α4 as well as the left end of β2. Actin binds to residues from β5, β6, β7, and α4. Two positively charged patches on the surface of profilin-II are candidate binding sites for PIP_2. One patch consists of residues from the right side of α3, β5, and β6. The other patch consists of residues from the β1/β2 loop, the α2/α3 loop, the left end of β6, the left end of β7, and the β7/β8 loop

Profilins bind to actin, poly-L-proline, and polyphosphoinositides (Fig. 2), each with distinct, but partially overlapping contacts on the surface of the protein (Fig. 1). Poly-L-proline binds to a hydrophobic groove between the N- and C-terminal helices (Archer et al. 1994; Metzler et al. 1994). Actin binds to residues of the C-terminal helix and the exposed right side of the central β-sheet (Schutt et al. 1993). Two positively charged surface patches are strong candidates for binding polyphosphoinositides, because their electrostatic surface potentials are greater in amoeba profilin-II, which binds lipids, than profilin-I, which does not (Fedorov et al. 1994a). The candidate binding site on the right side of the exposed β-sheet overlaps the actin binding site. The other candidate binding site is formed by residues from the loops between the elements of secondary structure on the left side of the profilin molecule, as illustrated.

The interaction of profilin with actin has the following consequences. Profilin binds to the barbed end of actin monomers (Vandekerckhove et al. 1989; Schutt et al.

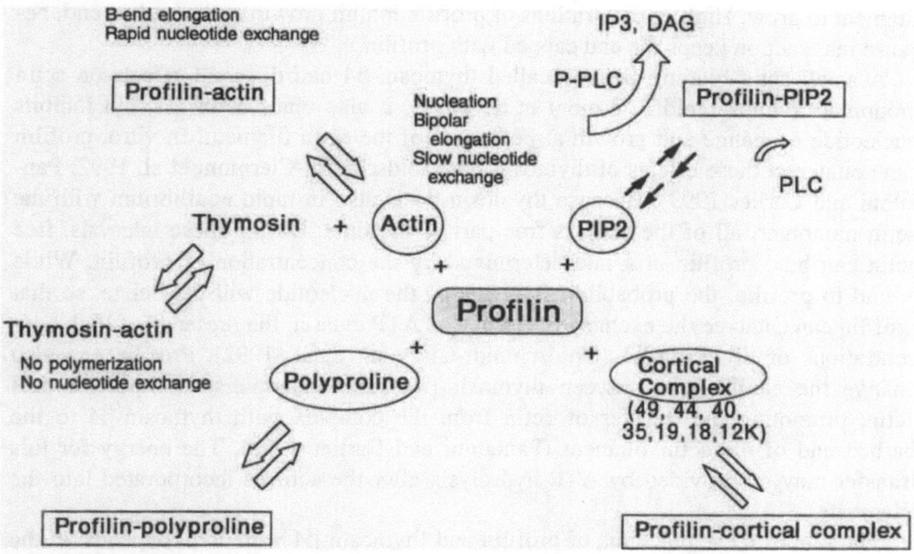

Fig. 2. A diagrammatic summary of the possible interactions of profilin with its various ligands. The *size of the arrows* is roughly proportional to the values of the rate constants for the reactions. The various ligands will compete for each other, depending upon mass action and the availability of profilin released from membrane binding sites by phospholipase C (PLC)

1993) inhibiting polymer nucleation and growth at the pointed end of actin filaments much more strongly than growth at the barbed end (Pollard and Cooper 1984). Dissociation of the complex of human profilin with actin is rapid (dissociation rate constant $k = 3\ s^{-1}$; Goldschmidt-Clermont et al. 1991b), so binding to actin is rapid and reversible at steady state. During this transient interaction, profilin catalyzes the exchange of the divalent cation and nucleotide bound to actin (Mockrin and Korn 1980; Nishida 1985; Goldschmidt-Clermont et al. 1991b). The dissociation rate constant for ATP is $0.01\ s^{-1}$ from actin alone and about $10\ s^{-1}$ from the actin-profilin complex. The most likely mechanism of dissociation is an opening of the cleft containing the nucleotide in the middle of the actin molecule. This change in the actin conformation is predicted by a theoretical, normal mode analysis of profilin dynamics (Tirion and ben-Avraham 1993), but an open conformation was not seen in the crystal structure of the complex (Schutt et al. 1993). The opening promoted by profilin may be a transient event or may be restricted by molecular contacts in the crystal.

Profilins have complex effects on actin polymerization. Profilins strongly inhibit nucleation of actin filaments and growth of the filaments at the slow growing, barbed end (Pollard and Cooper 1984), because the profilin sterically blocks the barbed end of the actin molecule (Schutt et al. 1993) preventing the association of the complex with the pointed end of other subunits and with the pointed end of the filament. In contrast, the complex of profilin with actin adds to the fast-growing, barbed end of actin filaments, because the pointed end of the complex is exposed. Once bound, profilin caps the barbed end of a filament, but profilin rapidly dissociates, allowing the filament to grow. High concentrations of profilin inhibit growth at the barbed end, because mass action keeps the end capped with profilin.

A small actin binding peptide called thymosin-β4 had different effects on actin monomers (Goldschmidt-Clermont et al. 1992). It also binds actin 1:1, but inhibits nucleotide exchange and growth at both ends of the actin filament. In vitro, profilin can counteract these effects of thymosin-β4 (Goldschmidt-Clermont et al. 1992; Pantaloni and Carlier 1993). Because thymosin-β4 is also in rapid equilibrium with the actin monomer, all of the actin is free part of the time. During these intervals, free actin can bind profilin at a rate determined by the concentration of profilin. While bound to profilin, the probability is high that the nucleotide will dissociate, so that profilin can catalyze the exchange of ADP for ATP even in the presence of high concentrations of thymosin-β4 (Goldschmidt-Clermont et al. 1992). Profilin can also change the equilibrium between thymosin-β4, actin monomers, and polymerized actin, promoting the transfer of actin from the complex with thymosin-β4 to the barbed end of the actin filament (Pantaloni and Carlier 1993). The energy for this transfer may be provided by ATP hydrolysis after the actin is incorporated into the filaments.

The rate of these reactions of profilin and thymosin-β4 with actin depends on the concentration of profilin, so it may be important for the cell to control the free profilin concentration. A candidate mechanism is profilin binding to polyphosphoinositides and the dissociation of the complex by enzymatic hydrolysis of the lipids.

Vertebrate profilins (Lassing and Lindberg 1985, 1988; Goldschmidt-Clermont et al. 1990; Machesky et al. 1990), amoeba profilin-II (Machesky et al. 1990), and vaccinia virus profilin (Machesky et al. 1994b) bind with micromolar affinity to small clusters of phosphatidylinositol 4-phosphate (PIP) and phosphatidylinositol 4,5-bisphosphate (PIP_2) in lipid bilayers. The association (10^4 M^{-1} s^{-1}) and dissociation (10^{-2} s^{-1}) rate constants are small, so the interaction of profilin with lipids is much less dynamic than its interaction with actin (Machesky et al. 1994b). The basic residues thought to be required for lipid binding are located in two patches, one of which overlaps the actin binding site (Fedorov et al. 1994a), explaining why lipid competes with actin for binding to profilin (Lassing and Lindberg 1985, 1988; Goldschmidt-Clermont et al. 1991b).

By binding PIP_2, profilin inhibits the production of the second messenger inositol trisphosphate (IP_3) and diacylglycerol (DAG) by soluble phosphoinositide-specific phospholipase Cs (Goldschmidt-Clermont et al. 1990; Machesky et al. 1990). Phosphorylation of phospholipase Cγ1 by activated EGF receptor tyrosine kinase overcomes the inhibition by profilin (Goldschmidt-Clermont et al. 1991a). Profilin inhibition and tyrosine kinase activation provide a biochemical mechanism to link growth factor stimulation to cellular production of IP_3. This biochemical mechanism still needs to be tested in vivo. Profilin does not inhibit all phosphoinositide-specific phospholipase Cs. For example, PLCβ, a phospholipase activated by a G protein rather than tyrosine phosphorylation (Taylor et al. 1991), is fully active in the presence of profilin (Goldschmidt-Clermont et al. 1991a).

Profilins also bind to polymers of L-proline (Tanaka and Shibata 1985). Multidimensional NMR revealed that both amoeba and human profilins bind poly-L-proline via aromatic and hydrophobic side chains exposed between the N- and C-terminal helices (Archer et al. 1994; Metzler et al. 1994). The site includes the conserved tryptophan-2. Mutagenesis of these residues inhibits binding to poly-L-proline (Bjorkegren et al. 1993) and vaccinia profilin, which lacks several of these residues, does not bind poly-L-proline (Machesky et al. 1994b). The affinity of profilins for poly-L-proline is relatively low, with a dissociation equilibrium constant of about 10^{-5} M under physiological conditions. By NMR in pure water the exchange rate is very rapid, with a dissociation rate constant of > 10^3 s^{-1} and a diffusion limited association rate constant of about 10^7 M^{-1} s^{-1} (Archer et al. 1994). We expect that when naturally occuring, proline-rich ligands for this site are found, they will have a higher affinity.

Acanthamoeba profilin binds a complex consisting of seven polypeptides with molecular weights of 49, 44, 40, 35, 19, 18, and 15 kDa (Machesky et al. 1994a). Several of these polypeptides colocalize in the cortex of the cell. The 49- and 44-kDa proteins are unconventional actins with low (< 50%) identity to conventional actin. They are similar to the actin-related proteins originally discovered in yeast (Lees-Miller et al. 1992; Schwob and Martin 1992). The complex of seven proteins binds profilin more strongly than actin, but is present in a much lower concentration, perhaps on the order of 1% the actin concentration. We are interested in the possibility that this complex may participate in the regulation of actin filament assembly in the cortex of the cell, but nothing yet is known about the functional properties of the components, since they have not been purified from any cell.

The interesting variety of profilin ligands illustrated in Fig. 2 raises the possibility that profilin may be a protein second messenger, carrying information from the phosphoinositide signaling pathways to the cytoskeleton via actin and to other systems via proline-rich proteins. At the present time, we have no direct evidence for this hypothesis, but the positive effect of flux through both tyrosine kinase-coupled and G-protein-coupled phosphoinositide signaling pathways on the assembly of cortical actin filaments (Dadabay et al. 1991) might be explained by the release of profilin from membrane phospholipids. The connection has not been confirmed at the cellular level, since assays are not yet available for either membrane association of profilin or the transient association of profilin with actin in the cytoplasm. The proline-rich target for profilin and improved methods for following the intracellular movements of profilin will be necessary to test this hypothesis.

Acknowledgments. The work from the author's laboratory was supported by research grants from the National Institutes of Health. The following individuals contributed to this work: S. Almo, C. Ampe, S. Archer, S. Atkinson, J.A. Cooper, P. Goldschmidt-Clermont, D. Kaiser, J. Kelleher, E. Lattman, L. Machesky, G. Petrella, D.L. Rimm, D. Torchia, J. Vandekerckhove, and V. Vinson. Their published work is cited in the references.

References

Archer SJ, Vinson VK, Pollard TD & Torchia DA (1993) Secondary structure and topology of *Acanthamoeba* profilin-I as determined by heteronuclear magnetic resonance spectroscopy. Biochemistry 32:6680–6687

Archer SJ, Vinson VK, Pollard TD & Torchia DA (1994) Elucidation of the poly-L-proline binding site in *Acanthamoeba*, profilin-I by NMR spectroscopy. FEBS Lett 337:145–151

Binette F, Benard M, Laroche A, Pierron G, Lemieux G & Pallotta D (1990) Cell-specific expression of a profilin gene family DNA Cell Biol 9:323–334

Bjorkegren C, Rozycki M, Schutt CE, Lindberg U & Karlsson R (1993) Mutagenesis of human profilin locates its poly (L-proline) binding site to a hydrophobic patch of aromatic amino acids. FEBS Letters 333:123–126

Carlsson L, Nystrom LE, Sundkvist L, Markey F & Lindberg U (1977) Actin polymerizability is influenced by profilin, a low molecular weight protein in non-muscle cells. J Mol Biol 115:465–483

Cooley L, Verheyen E & Ayers K (1992) Chickadee encodes a profilin required for intercellular cytoplasmic transport during *Drosophila* oogenesis. Cell 69:173–184

Dadabay CY, Patton E, Cooper JA & Pike LJ (1991) Lack of correlation between changes in polyphosphoinositide levels and actin/gelsolin complexes in A431 cells treated with epidermal growth factor. J Cell Biol 112:1151–1156

Fedorov AA, Lattman EE, Pollard TD & Almo SC (1995) Comparison of the X-ray structures of human and amoeba profilins. in preparation

Fedorov AA, Magnus KA, Graupe H, Lattman EE, Pollard TD & Almo SC (1995) X-ray structures of isoforms of the actin binding protein profilin that differ in their affinity for polyphosphoinositides. Proc Nat Acad Sci USA 91:8636–8640

Goldschmidt-Clermont PJ, Furman MI, Wachsstock DH, Safer D, Nachmias VT & Pollard TD (1992) Regulation of actin nucleotide exchange by thymosin β4 and profilin. Molec Biol Cell 3:1015–1024

Goldschmidt-Clermont PJ, Kim JW, Machesky LM, Rhee SG & Pollard TD (1991a) Regulation of phospholipase C-γl by profilin and tyrosine phosphorylation. Science 251:1231–1233

Goldschmidt-Clermont PJ, Machesky LM, Baldassare JJ & Pollard TD (1990) The actin-binding protein profilin binds to PIP_2 and inhibits its hydrolysis by phospholipase C. Science 247:1575–1578

Goldschmidt-Clermont PJ, Machesky LM, Doberstein SK & Pollard TD (1991b) Mechanism of the interaction of human platelet profilin with actin. J Cell Biol 113:1081–1089

Haarer BK, Lillie SH, Adams AEM, Magdolen V, Bandlow W & Brown SS (1990) Purification of profilin from *Saccharomyces cerevisiae* and analysis of profilin-deficient cells. J Cell Biol 110:105–114

Haugwitz M, Noegel AA, Rieger D, Lottspeich F & Schleicher M (1991) *Dictyostelium* contains two profilin isoforms that differ in structure and function. J Cell Sci 100:481–489

Honore B, Madsen P, Andersen AH and Leffers H (1993) Cloning and expression of a novel human profilin variant, profilin-II. FEBS Lett 330:151–155

Kaiser DP, Sato MD, Ebert R & Pollard TD (1986) Purification and characterization of two isoforms of *Acanthamoeba* profilin. J Cell Biol 102:221–226

Lassing I & Lindberg U (1985) Specific interaction between phosphatidylinositol 4,5 bisphosphate and profilactin. Nature 318:472–474

Lassing I & Lindberg U (1988) Specificity of the interaction between PIP_2 and the profilin:actin complex. J Cell Biochem 37:255–268

Lees-Miller JP, Henry G & Helfman DM (1992) Identification of *act2*, an essential gene in the fission yeast *Schizosaccharomyces pombe* that encodes a protein related to actin. Proc Nat Acad Sci USA 89:80–83

Machesky LM & Pollard TD (1993) Profilin as a potential mediator of membrane cytoskeletal communications. Trends Cell Biol 3:381–385

Machesky LM, Ampe C, Atkinson SJ, Vandekerckhove J & Pollard TD (1994a) A cortical complex of seven *Acanthamoeba* polypeptides including two unconventional actin binds to profilin. J Cell Biol 127:107–115

Machesky LM, Cole NB, Moss B & Pollard TD (1994b) Vaccinia virus expresses a novel profilin with a higher affinity for polyphosphoinositides than actin. Biochemistry 33:10815–10824

Machesky LM, Goldschmidt-Clermont PJ & Pollard TD (1990) The affinities of human platelet and *Acanthamoeba* profilin isoforms for polyphosphoinositides account for the relative abilities to inhibit phospholipase C. Cell Regulation 1:937–950

Metzler WJ, Bell AJ, Ernst E, Lavoie TB & Mueller L (1994) Identification of the poly-L-proline binding site on human profilin. J Biol Chem 269:4620–4625

Metzler WJ, Constantine KL, Friedrichs MS, Bell AJ, Ernst EG, Lavoie TB & Mueller L (1993) Characterization of the three-dimensional structure of human profilin: ^{1}H, ^{13}C and ^{15}N NMR assignments and global folding pattern. Biochemistry 32:13818–13829

Mockrin SC & Korn ED (1980) *Acanthamoeba* profilin interacts with G-actin to increase the exchange of actin bound adenosine 5'-triphosphate. Biochemistry 19:5359–5362

Nishida E (1985) Opposite effects of profilin and cofilin from porcine brain on the rate of exchange of actin-bound ATP. Biochemistry 24:1160–1164

Pollard TD & Quirk S (1994) Profilins, ancient actin binding proteins with highly divergent primary structures. In: "Molecular Evolution of Physiological Processes" Ed by Fambrough D. Rockefeller University Press, p 117–128

Pollard TD & Rimm DL (1991) Analysis of cDNA clones for *Acanthamoeba* profilin-I and profilin-II shows end to end homology with vertebrate profilins and a family of profilin genes. Cell Motil Cytoskel 20:169–177

Pollard TD & Cooper JA (1984) Quantitative analysis of the effect of *Acanthamoeba* profilin on actin filament nucleation and elongation. Biochemistry 23:6631–6641

Schutt C, Myslik JC, Rozycki MD, Goonesekere NCW & Lindberg U (1993) The structure of crystalline profilin-β-actin. Nature 365:810–816

Schwob E & Martin R (1992) New yeast actin-like gene required late in the cell cycle. Nature 355:179–182

Tanaka M & Shibata H (1985) Poly(L-proline)-binding proteins from chick embryos are a profilin and profilactin. European J Biochem 151:291–297

Taylor SJ, Chae HZ, Rhee SG & Exton JH (1991) Activation of the β1 isozyme of phospholipase C by purified alpha subunits of the G_q class of G proteins. Nature 350:516–518

Tirion MM & Ben-Avraham D (1993) A normal mode analysis of G-actin. J Mol Biol 230:186–195

Valenta R, Duchene M, Pettenburger K, Sillaber C, Valent P, Bettelheim P, Breitenbach M, Rumpold H, Kraft D & Scheiner O (1991) Identification of profilin as a novel pollen allergen; IgE autoreactivity in sensitized individuals. Science 253:557–560

Vandekerckhove JS, Kaiser DA & Pollard TD (1989) *Acanthamoeba* actin and profilin can be crosslinked between glutamic acid 364 of actin and lysine 115 of profilin. J Cell Biol 109:619–626

Vinson VK, Archer SJ, Lattman EE, Pollard TD & Torchia DA (1993) Three-dimensional solution structure of *Acanthamoeba* profilin-I. J Cell Biol 122:1277–1283

Membrane-Microfilament Attachment Sites: the Art of Contact Formation

B. M. JOCKUSCH, M. KROEMKER, and K. SCHLÜTER[1]

1 Introduction

In animals, embryogenesis and wound healing require a finely tuned coupling of signal reception with the intracellular cytoskeleton. In particular, tissue formation is specified by the differentiation of discrete cellular contact sites between cells or between cells and their substratum. These contacts are morphologically distinct structures, but are also highly dynamic, and their reversible assembly is controlled by intra- as well as by extracellular factors. Regarding the cytoplasmic site of these contacts, the data from numerous studies imply that there is a catalogue of structural proteins linking the distal portions of the microfilaments to the integral components of the plasma membrane. The interaction of these structural proteins with each other is probably modulated by posttranslational modification and by accessory proteins mediating between the cytoskeletal elements and the signaling pathways. Thus, forming these contacts must be a highly complex event. Only a detailed analysis of each of the protein modules involved in such interactions, their posttranslational modification, and their "cross talk" will provide the basis for the understanding of the entire process. Table 1 gives a summary of the major proteins found in adhesive contacts of the cell-substratum type (focal contacts) and their classification according to their proposed function, as described in numerous articles (reviewed in Burridge et al. 1988; Geiger and Ginsberg 1991; Turner and Burridge 1991; Critchley et al. 1991; Luna and Hitt 1992; Schaller and Parsons 1993). Classical examples of structural components are, of course, actin itself, α actinin, and vinculin. In addition, talin and radixin, two proteins with sequence homology to band 4.1, an actin-membrane-linking protein of the red blood cell, fall into this category. Regulatory proteins comprise the profilins and a group of Zn finger proteins (LIM proteins, zyxin, paxillin, and the cysteine-rich protein, CRP), and the proline-rich protein VASP (vasodilator stimulated phosphoprotein) which was originally discovered as a major substrate for ser/thr kinases in platelets but has been found as a general component in focal contacts of many different cell types (Reinhard et al. 1992). In addition, several tyrosine and serine/threonine specific kinases have been identified there, in particular the focal adhesion kinase FAK. Among the proteins which probably display both structural as well as regulatory components are tensin, a large multidomain protein, and various integral membrane receptors (see below). For cell-cell contacts, as exemplified in the epithelial zonula adherens junctions, such a list would have to be slightly modified.

---

[1] Cell Biology, Institute of Zoology, Technical University of Braunschweig, 38092 Braunschweig, FRG.

45. Colloquium Mosbach 1994
The Cytoskeleton

Table 1. Protein components identified in microfilament-plasma membrane attachment sites of the focal contact type

Structural proteins	Regulatory proteins	Dual function proteins
Actin	Profilin	Tensin
Alpha actinin	Zyxin	Growth factor receptors
Vinculin	Cysteine-rich protein	ECM[b] receptors
Talin	VASP[a]	
Radixin	Tyrosine-kinases Ser/Thr-kinases	

[a] VASP: vasodilator stimulated phosphoprotein.
[b] ECM: extracellular matrix.

For example, talin is missing from these sites (Geiger et al. 1985), while in addition to vinculin, a vinculin relative, α catenin, and several additional proteins have been identified as part of a structural link to the epithelial cell adhesion molecules (Herrenknecht et al. 1991; Nagafuchi et al. 1991).

Three novel aspects have emerged from the most recent studies on microfilament-plasma membrane linkage: (1) actin monomers are provided for the site-directed growth of filaments from membrane bound complexes; (2) there is certainly not a single, linear chain of molecules leading from the end of actin filaments through integral membrane proteins towards the extracellular matrix (ECM) or an adjacent cell, but there are multiple ways in which the structural proteins can interact; (3) many of the proteins involved seem to perform both structural and regulatory functions. For example, several of the integral membrane proteins serving as receptors for components of the extracellular matrix (integrins) or growth factors not only bind directly to structural microfilament proteins (Burridge et al. 1988; Den Hartigh et al. 1992; Van Etten et al. 1994) but are also part of an intracellular signaling machinery which modifies a large number of microfilament proteins (Damsky and Werb 1992; Schaller and Parsons 1993; Lo and Chen 1994).

In this chapter, we want to illustrate some general principles of building cell-substratum contacts. To this end, we will describe structural and functional aspects of two proteins, vinculin and profilin, that may serve as examples for a structural and a regulatory component, respectively. In this context we will describe some of our own work designed to contribute to the understanding of structure and function of adhesive junctions.

2 Vinculin Is a Multiligand Protein with Unique Structural Properties

This 116-kDa protein is concentrated at the cytoplasmic face of cell-cell and cell-substratum junctions in a wide variety of organisms and tissues (Geiger 1979), ranging from man to nematodes. In vertebrate and invertebrate muscle, it is present at the anchorage sites between myofibrils and the muscle membrane (Pardo et al. 1983; Lin et al. 1989; Barstead and Waterston 1991; Jockusch et al. 1993). In various adhesive nonmuscle cells, such as fibroblasts, vinculin seems indispensible for cellular attachment and normal mobility. This has been established by several independent methods: interfering with the correct vinculin level by overexpression suppresses cellular locomotion and enhances cell attachment (Fernández et al. 1993), while vinculin antisense RNA transfection shows the opposite effect (Fernández et al. 1992). In addition, we demonstrated that microinjection of monoclonal antibodies against specific vinculin epitopes disrupts focal contacts (Westmeyer et al. 1992). From such studies, it has been concluded that vinculin is directly involved in establishing the prominent focal contacts seen in cultured adhesive cells and their substratum. Conversely, it has also been shown that the type of substratum influences vinculin synthesis in fibroblasts (Ungar et al. 1986; Bendori et al. 1987). Interestingly, vinculin is not only confined to the large, stable focal contacts of tissue forming cells, but is also located at membrane-microfilament attachment sites in single, vagabonding cells of the blood system and of platelets (Rosenfeld et al. 1985; Nachmias and Golla 1991). These cells form much smaller and less permanent contacts, which are, however, also morphologically discrete structures. Interestingly, no vinculin-like polypeptide has yet been discovered in protozoan and slime mold amoebae, cells which seem continuously on the move.

Despite these numerous data on occurrence, expression, and localization of vinculin, its precise role in linking microfilaments to the integral membrane proteins of the contact site is still elusive. The molecule, as isolated from smooth muscle, is monomeric, and contains a globular (N-terminal) head domain of approximately 85 kDa linked to a 29-kDa rod-like tail. When spread onto mica, many of the molecules display the tail as an extended structure (Milam 1985; Molony and Burridge 1985), and we could show that this is also the preferred shape in solution (Eimer et al. 1993). Head and tail domains can be conveniently separated by proteolysis and conventional gel chromatography, so that the properties of both fragments can be analyzed independently. Binding studies with the intact molecule, proteolytic fragments, and fragments expressed as recombinant fusion proteins have yielded the following information: in vitro, vinculin binds to tensin (Wilkins et al. 1987), talin (Jones et al. 1989), and α actinin (Wachsstock et al. 1987; Critchley et al. 1991). The isolated head binds to talin (Groesch and Otto 1990; Johnson and Craig 1994), α actinin (Kroemker et al. 1994) and to itself (Eimer et al. 1993), while the tail interacts with filamentous actin (Jockusch and Isenberg 1981, 1982; Isenberg et al. 1982; Ruhnau and Wegner 1988; Menkel et al. 1994), with the regulatory protein paxillin (Turner et al. 1990; Wood et al. 1994), and also with itself (Molony and Burridge 1985; Eimer, Niermann, Kroemker, Rüdiger, and Jockusch, unpub.). In addition, vinculin has been shown to interact directly with acidic phospholipids (Ito et al. 1983; Isenberg 1991) which are also mediated by the tail fragment (cf. Isenberg and Goldmann 1992; Johnson and Craig 1994). By expressing vinculin fragments in cells, by microinjection, and by incubating vinculin and its fragments with detergent-extracted cell models, it was shown that the intact molecule as well as its head fragment target faithfully to the end of microfilament bundles at the focal contact site (Ball et al. 1986; Burridge and Feramisco 1980; Menkel et al. 1994). This is also demonstrated in Fig. 1. These data are consistent with the hypothesis that vinculin is directed towards these

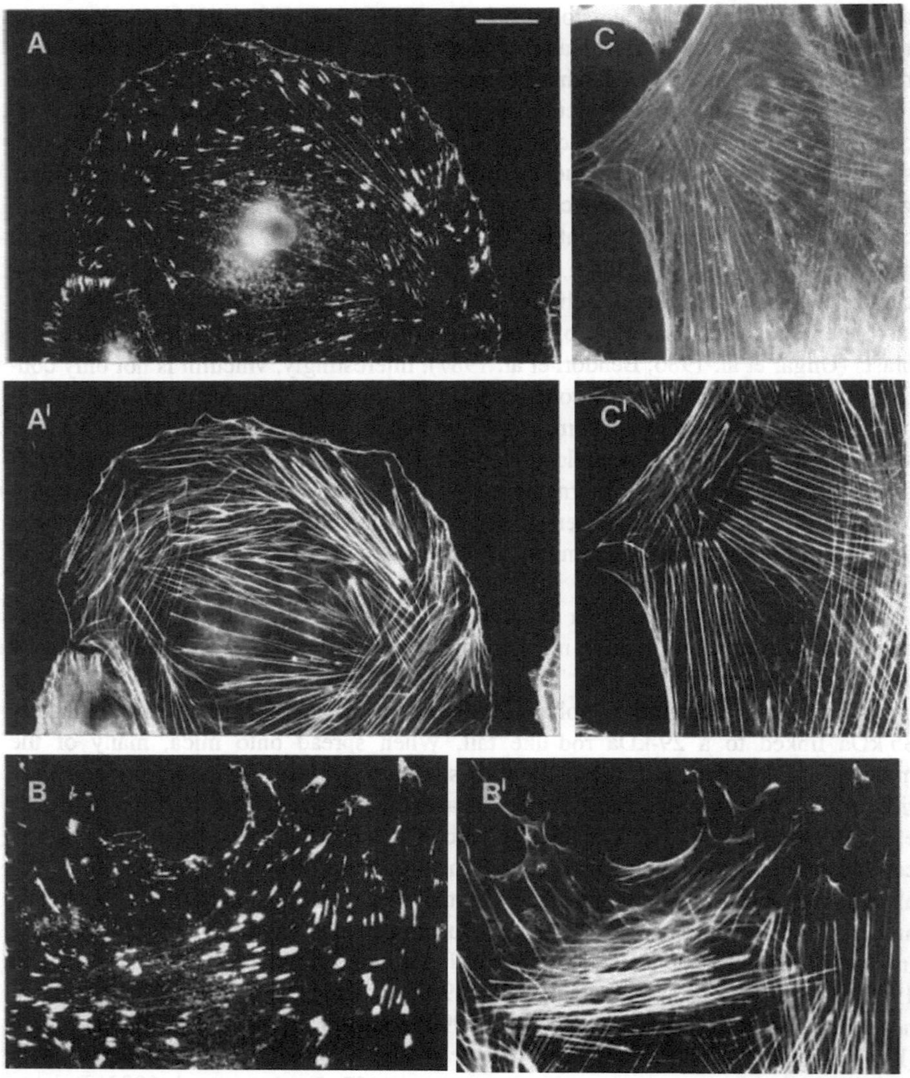

Fig. 1 A–C. Targeting of chicken gizzard vinculin and its proteolytic fragments to microfilament bundles in cell models. SW 3T3 fibroblasts were fixed and extracted with detergent and incubated with the various proteins. Double fluorescence images were obtained by labeling gizzard vinculin (**A**) and its 85-kDa head piece (comprising amino acid residues 1–850, **B**) with chicken-specific monoclonal antibodies directed to an epitope in the vinculin head and rhodamine-conjugated second antibodies. The tail fragment of gizzard vinculin (amino acid residues 857–1066) was directly conjugated with rhodamine (**C**). Filamentous actin was revealed by staining with FITC-phalloidin (**A'**, **B'**, **C'**). *Bar* 20 μm. While the intact vinculin and the head piece concentrate exclusively at the ends of microfilament bundles (focal contact area, **A** and **B**), the isolated tail fragments decorate the microfilament bundles along their entire length (**C**), demonstrating the existence of a functional actin binding-domain in the tail piece

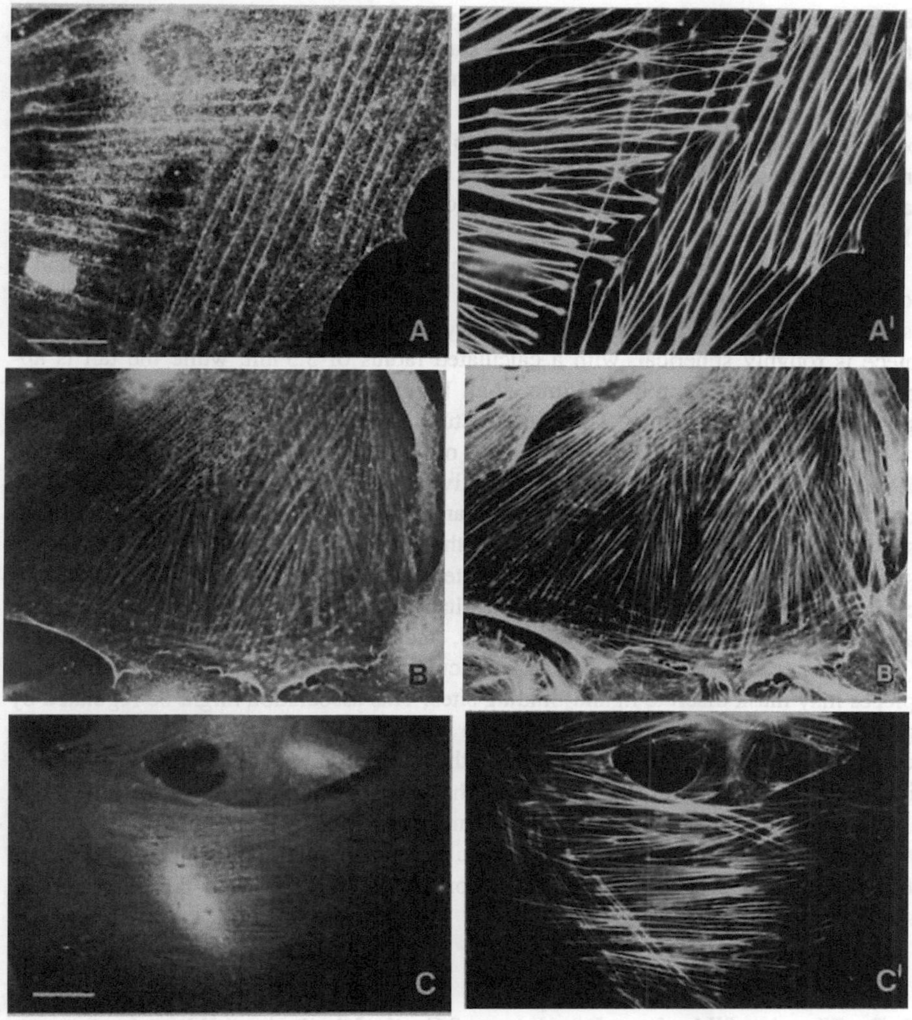

Fig. 2 A–C. Targeting of recombinant vinculin fragments to microfilament bundles in cell models. Cells and conditions were as described in Fig. 1. The recombinant proteins were chimeric proteins of the maltose binding protein (MBP) of *E. coli* fused to vinculin sequence stretches comprising amino acid residues 808–1063 (**A**), 893–1066 (**B**), and 1016-1066 (**C**), their location was revealed by an MBP antibody followed by a second antibody conjugated to rhodamine. The corresponding images of filamentous actin are seen in **A'**, **B'**, and **C'**. *Bars* 20 µm. While fusion proteins comprising the entire vinculin tail and a shorter fragment (residues 893–1066) both decorate the microfilament bundles quite similar to the genuine fragment (cf. Fig. 1 C with **A** and **B**), a fusion protein harboring only the 50 most carboxyterminal residues of vinculin fail to do so (**C**). Combining the data from these results with those of Fig. 1 allows for the conclusion that the actin binding domain resides within residues 893–1016 of the vinculin tail

regions by its affinity for talin, as the affinity between vinculin and talin is fairly high (the K_d has been determined to 10^{-8} M, Burridge and Mangeat 1984), while for the other ligands known, it is generally in the micromolar range (cf. Menkel et al. 1994; McGregor et al. 1994). On the other hand, isolated tail pieces and tail fragments as part of a recombinant fusion protein bind to filamentous actin in cells, supporting the in vitro studies (Menkel et al. 1994; Figs. 1 and 2). By using deletion mutants of the tail, we could show that this actin binding domain resides within residues 893–1016 (Fig. 2).

While such studies allow for a mapping of functional domains in vinculin and a topographic assignment within the vinculin sequence, our knowledge on its structure is very limited. As determined by photon correlation spectroscopy, the isolated head piece is roughly spherical, with a calculated radius of 3.3 nm, while the tail is approximately 13 to 16 nm long (Eimer et al. 1993; Eimer, Niermann, Kroemker, Rüdiger, and Jockusch, unpub.). The two structural domains are connected by a proline-rich hinge which confers a certain amount of flexibility to the molecule (Eimer et al. 1993). This flexibility allows for an attractive model on the regulation of the interaction between vinculin and its numerous ligands: The position of the tail piece with respect to the entire particle may regulate the binding of different ligands including self-association. Such a concept is supported by data showing that the isolated vinculin head has a higher affinity for talin than the intact vinculin molecule (Groesch and Otto 1990; Johnson and Craig 1994), and the same is observed for α actinin (Kroemker et al. 1994). Thus, an intramolecular association between the head and tail pieces may mask discrete ligand-binding sites. This association might be regulated by phosphorylation of the tail piece (Johnson and Craig 1994). Accordingly, intramolecular mobility of vinculin could influence the interaction with other components within the adhesive junction. In the case of the focal contact, expanded vinculin molecules, targeted to the inner face of the plasma membrane by the affinity of their heads to talin and self-assembled into oligomeric structures, might collect the ends of actin filament bundles via the actin-binding domain of their tails into the characteristic spear-tip structures.

3 Profilins Are Ubiquitous Regulators of Cortical Actin Assembly and Microfilament-Membrane Attachment

The rapid and regulated protrusion of the plasma membrane, as seen in cell spreading and anchorage, as well as in ruffling and locomotion, requires the provision of polymerization-competent actin subunits from a precursor pool. Therefore, membrane-associated proteins capable of a signal-regulated complex formation with monomeric actin should play an important role in microfilament-membrane linkage. Profilin, a small protein originally described as an inhibitor of actin polymerization (Carlsson et al. 1977), seems to meet with these requirements (Haarer and Brown 1990; Theriot and Mitchison 1993; Pantaloni and Carlier 1993). It has turned out that the originally described protein belongs to a large family whose members are ubiquitously found in yeasts, slime molds, plants, invertebrates, and vertebrates, and even in Vaccinia virus

(Machesky and Pollard 1993). Profilins are characterized by their ability to form complexes with each actin, and also to interact with phosphatidylinositol-4,5-bisphosphate (PIP_2, Lassing and Lindberg 1988) at the plasma membrane which links them to several signal transduction pathways. Additionally, they bind to poly-L-proline which is conveniently used to purify the protein from tissue, or, as a recombinant product, from bacterial extracts. It is conceivable that in cells, this poly-L-proline binding activity endows profilins to interact with proline-rich proteins also engaged in microfilament organization. Sequence analyses have shown that different profilins display surprisingly little overall homology, but the recently published NMR and X-ray high resolution data (Vinson et al. 1993; Schutt et al. 1993) demonstrate that they share a rather highly conserved structure (Rozycki et al. 1994). This is also illustrated in Fig. 3 which shows a structural model of birch pollen profilin, derived from the primary sequence data and fitted into the atomic structure of bovine profilin obtained from profilin-actin crystals (Schutt et al. 1993).

Apart from yeast, all species seem to have at least two profilins, and plants express an even higher number. In some organisms, a tissue- and developmental stage-regulated synthesis of profilins has been found (Buss and Jockusch 1989; Haugwitz et al. 1991; Cooley et al. 1992; Staiger et al. 1993). On the other hand, the basic profilin purified from many animal cell lines and tissues forms complexes with both nonmuscle actins, the β and γ isoforms (Segura and Lindberg 1984). It seems conceivable that in cells there is a spatial separation between these moieties: highly dynamic areas, such as the leading lamellae and ruffles, are especially active in β-actin synthesis (Latham et al. 1994; Hill et al. 1994), and these regions coincide with a high profilin concentration (Buss et al. 1992). Therefore, in these areas, preferentially profilin-β-actin complexes may be localized in high concentrations, serving as actin filament precursors at membrane-attachment sites.

We have tackled the question of the degree of a functional substitution between profilin of low sequence homology. We examined the interaction of genuine and recombinant birch pollen profilin with muscle and nonmuscle actins, and compared the affinities of the various complexes obtained by determining their dissociation constants. In viscometric assays and crosslinking experiments, we could demonstrate a complex formation between the plant and mammalian proteins, and the dissociation constants obtained were in the micromolar range, as had been previously determined for homologous vertebrate complexes. Quite analogous to the homologous profilin, the heterologous plant-vertebrate profilin-actin complex could be isolated by gel chromatography, and the birch protein competed with thymosin β4, another small G-actin binding protein, for the same binding site on actin (Giehl et al. 1994). These data show that the actin-binding domains in plant and animal profilins are functionally highly conserved, although the overall sequence similarity is less than 25%.

Based on in vitro results, we have constructed two mammalian cell lines stably expressing the plant protein. The BHK-21 clones obtained differ in their birch profilin content, but in neither case was the actin level, the F-actin organization nor the level of endogeneous profilin altered. The overall profilin:actin ratios were 1:2 and 1:3. With a birch profilin-specific antibody, we could show that the plant protein was concentrated in highly dynamic areas of the cells, again similar to its endogeneous counterpart. Ruffling areas of the membrane, as well as the ends of microfilament bundles

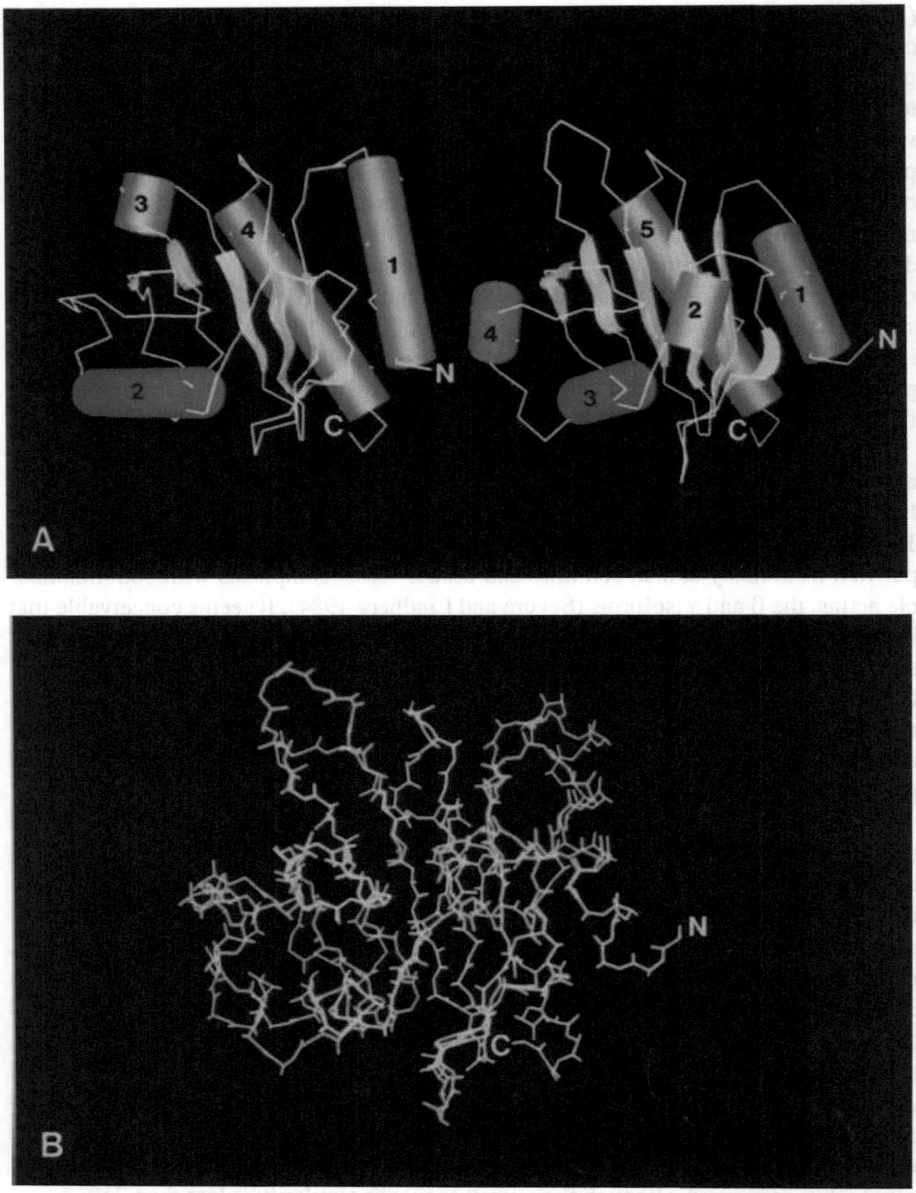
3
4
1
2
N
C
A
5
2
1
4
3
N
C
N
C
B

within the focal contact sites were especially enriched in the plant protein, confirming previous data we had obtained with microinjecting the birch profilin (Rothkegel et al. 1994). With respect to cell attachment, motility, and proliferation, these clones were indistinguishable from the parental BHK-21 line. However, we found that the microfilament system in the transfectants was more stable against cytochalasin D-induced breakdown as compared with controls (Rothkegel et al. 1994). A stabilizing effect of profilin overexpression on actin filaments has also been reported in CHO cells transfected with human profilin (Finkel et al. 1994). Thus, it seems that at least in this system, the plant protein can functionally substitute for its mammalian relative. Interestingly, in Dictyostelium mutants lacking both profilin isoforms, a similar stabilization effect on actin filaments was observed (Haugwitz et al. 1994). These results, seemingly at variance with each other, might reflect differences in actin filament regulation between slime mold amoebae and mammalian cells.

4 Conclusions

In this chapter, we have outlined the spatial organization and regulation of microfilament attachment to the inner face of the plasma membrane. As examples for the numerous elements involved, we have described the proposed roles of a structural component, vinculin, and a regulatory protein, profilin, in this process. Clearly, many more studies are needed to understand every aspect of the contact formation, but some general rules seem to emerge from what is known today. The stunning complexity of the attachment sites, reflected in the numerous components localized there, is probably needed to ensure their dynamics: forming and breaking contacts between multiple ligands with moderate or weak affinities for each other is easier to regulate than molecular links involving only a few members. Spatial and temporal regulation is again highly complex. Provisions of subunits, conformational changes in structural components, competition between the numerous ligands for a given binding site and posttranslational modification may all be involved in selecting specific intermolecular bonds.

Fig. 3 A, B. Structural model of birch pollen profilin as obtained by homology modelling, using the high resolution data available for bovine profilin (Schutt et al. 1993) and the computer program BRAGI. **A** Secondary structures of the birch profilin model (*left*) and bovine profilin (*right*). The structures are quite similar: a central β sheet of six strands is flanked by the first and the last α helix on one side, and further α helices on the other side. The first and the last helices of both proteins are identical, helix 2 of birch profilin is corresponding to helix 3 of bovine profilin. Helix 2 of bovine profilin and helix 3 of birch profilin are uncommon and not α-helical. At the position corresponding to the fourth helix in bovine profilin, the program does not recognize an α helix in the plant protein, because the bond lengths and angles are outside the range of tolerance. **B** Stick model of the superimposed backbones of birch (*green*) and bovine (*orange*) profilin, shown in the same orientation as in **A**. Overlapping parts appear in *grey* and reveal the degree of similarity (Collaboration with Drs. U. Lessel and D. Schomburg, GBF, Braunschweig, FRG)

Acknowledgments. We thank Drs. U. Lindberg and C. E. Schutt for providing the coordinates of bovine profilin prior to their listing in a data bank, and Dr. M. Rüdiger for critical reading of the manuscript. Our own work described in this chapter includes results obtained by Drs. A. Menkel, M. Rothkegel, M. Rüdiger, as well as K. Giehl and P. Bubeck in our group. It was financially supported by the Deutsche Forschungsgemeinschaft.

References

Ball EH, Freitag C, Gurofsky S (1986) Vinculin interaction with permeabilized cells: disruption and reconstitution of a binding site. J Cell Biol 103:641–648

Barstead RJ, Waterston RH (1991) Vinculin is essential for muscle function in the nematode. J Cell Biol 114:715–724

Belkin AM, Koteliansky VE (1987) Interaction of iodinated vinculin, metavinculin and α actinin with cytoskeletal proteins. FEBS Lett 220:291–294

Bendori R, Salomon D, Geiger B (1987) Contact-dependent regulation of vinculin expression in cultured fibroblasts: a study with vinculin-specific cDNA probes. EMBO J 6:2897–2905

Bendori R, Salomon D, Geiger B (1989) Identification of two distinct functional domains of vinculin involved in its association with focal contacts. J Cell Biol 108:2383–2393

Burridge K, Feramisco J (1980) Microinjection and localization of a 130 K protein in living fibroblasts: a relationship to actin and fibronectin. Cell 19:587–595

Burridge K, Fath K, Kelly T, Nuckolls G, Turner CE (1988) Focal adhesions: transmembrane junctions between the extracellular matrix and the cytoskeleton. Annu Rev Cell Biol 4:487–525

Burridge K, Mangeat P (1984) An interaction between vinculin and talin. Nature 308:744–748

Buss F, Jockusch BM (1989) Tissue specific expression of profilin. FEBS Lett 249:31–34

Buss F, Temm-Grove CJ, Henning S, Jockusch BM (1992) Distribution of profilin in fibroblasts correlates with the presence of highly dynamic actin filaments. Cell Motil Cytoskeleton 22:51–61

Carlsson L, Nystrøm LE, Sundkvist I, Markey F, Lindberg U (1977) Actin polymerizability is influenced by profilin, a low molecular weight protein in nonmuscle cells. J Mol Biol 115:465–483

Cooley L, Verheyen E, Ayers K (1992) Chickadee encodes a profilin required for intercellular cytoplasm transport during *Drosophila* oogenesis. Cell 69:173–184

Critchley DR, Gilmore A, Hemmings L, Jackson P, McGregor A, Ohanian V, Patel B, Waites G, Wood C (1991) Cytoskeletal proteins in adherens-type cell-matrix junctions. Biochem Soc Transact 19:1028–1033

Damsky CH, Werb Z (1992) Signal transduction by integrin receptors for extracellular matrix: cooperative processing of extracellular information. Curr Op Cell Biol 4:772–781

Den Hartigh JC, van Bergen en Henegouwen PMP, Verkleij AJ, Boonstra J (1992) The EGF receptor is an actin-binding protein. J Cell Biol 119:349–355

Eimer W, Niermann M, Eppe MA, Jockusch BM (1993) Molecular shape of vinculin in aqueous solution. J Mol Biol 229:146–152

Fernández JLR, Geiger B, Salomon D, Ben-Ze'ev A (1992) Overexpression of vinculin suppresses cell motility in BALBc/c 3T3 cells. J Biol Chem 255:1194–1199

Fernández JLR, Geiger B, Salomon D, Ben-Ze'ev A (1993) Suppression of vinculin expression by antisense transfection confers changes in cell morphology, motility, and anchorage-dependent growth of 3T3 cell. J Cell Biol 122:1285–1294

Finkel T, Theriot JA, Dise KR, Tomaselli GF, Goldschmidt-Clermont P (1994) Dynamic actin structures stabilized by profilin. Proc Natl Acad Sci USA 91:1510–1514

Geiger B (1979) A 130 K protein from chicken gizzard: its localization at the termini of microfilament bundles in cultured chicken cells. Cell 18:193–205

Geiger B, Ginsberg D (1991) The cytoplasmic domain of adherens-type junctions. Cell Motil Cytoskeleton 20:1–6

Geiger B, Volk T, Volberg T (1985) Molecular heterogeneity of adherens junctions. J Cell Biol 101:1523–1531

Giehl K, Valenta R, Rothkegel M, Ronsiek M, Mannherz HG, Jockusch BM (1994) Interaction of plant profilin with mammalian actin. Eur J Biochem 226:681–689

Groesch ME, Otto JJ (1990) Purification and characterization of an 85 kDA talin-binding fragment of vinculin. Cell Motil Cytoskeleton 15:41–50

Haarer BK, Brown SS (1990) Structure and function of profilin. Cell Motil Cytoskeleton 17:71–74

Haugwitz M, Noegel AA, Rieger D, Lottspeich F, Schleicher M (1991) *Dictyostelium discoideum* contains two profilin isoforms that differ in structure and function. J Cell Sci 100:481–489

Haugwitz M, Noegel AA, Karakesisoglou J, Schleicher M (1994) *Dictyostelium* amoebae that lack G-actin sequestering profilins show defects in F-actin content, cytokinesis and development. Cell, in press

Herrenknecht K, Ozawa M, Ekerskorn C, Lottspeich F, Lenter M, Kemler R (1991) The uvomorulin-anchorage protein α catenin is a vinculin homologue. Proc Natl Acad Sci USA 88:9156–9160

Hill MA, Schedlich L, Gunning P (1994) Serum-induced signal transduction determines the peripheral location of β-actin mRNA within the cell. J Cell Biol 126:1221–1230

Isenberg G (1991) Actin binding proteins-lipid interactions. J Muscle Res Cell Mot 12:136–144

Isenberg G, Goldmann WH (1992) Actin-membrane coupling: a role for talin. J Muscle Res Cell Mot 13:587–589

Isenberg G, Leonard K, Jockusch BM (1982) Structural aspects of vinculin-actin interactions. J Mol Biol 158:231–249

Ito S, Werth DK, Richert ND, Pastan I (1983) Vinculin phosphorylation by src kinase. Interaction of vinculin with phospholipid vesicles. J Biol Chem 258:14626–14631

Jockusch BM, Isenberg G (1982) Vinculin and α actinin interaction with actin and effect on microfilament network formation. Cold Spring Harbor Symp Quant Biol 46:613–623

Jockusch BM, Isenberg G (1981) Interaction of α actinin and vinculin with actin: opposite effects on filament network formation. Proc Natl Acad Sci USA 78:3005–3009

Jockusch BM, Wiegand C, Temm-Grove CJ, Nikolai G (1993) Dynamic aspects of microfilament-membrane attachments. Soc Exp Biol 47:253–266

Johnson RP, Craig SW (1994) An intramolecular association between the head and tail domains of vinculin modulates tail binding. J Biol Chem 269:12611–12619

Jones P, Jackson P, Price GJ, Patel B, Ohanion V, Lear AL, Critchley D (1989) Identification of a talin-binding site in the cytoskeletal protein vinculin. J Cell Biol 109:2917–2927

Kroemker M, Rüdiger AH, Jockusch BM, Rüdiger M (1994) Intramolecular interactions in vinculin control α-actinin binding to the vinculin head. FEBS Lett 355:259–262

Lassing H, Lindberg U (1988) The effect of divalent cations on the interaction between calf spleen profilin and different actins. Biochim Biophys Acta 953:95–105

Latham Jr VM, Kislauskis EH, Singer RH, Ross AF (1994) β-actin mRNA localization is regulated by signal transduction mechanisms. J Cell Biol 126:1211–1220

Lin Z, Eshleman J, Grund C, Fischman DA, Masaki T, Franke WW, Holzer H (1989) Differential response of myofibrillar and cytoskeletal proteins in cells treated with phorbol myristate acetate. J Cell Biol 108:1079–1091

Lo SH, Chen LB (1994) Focal adhesion as a signal transduction organelle. Cancer and Metastasis Rev 13:9–24

Luna EJ, Hitt AL (1992) Cytoskeleton-plasma membrane interactions. Science 258:955–964

Machesky LM, Pollard TD (1993) Profilin as a potential mediator of membrane-cytoskeleton communication. Trends Cell Biol 3:381–385

McGregor A, Blanchard AD, Rowe AJ, Critchley DR (1994) Identification of the vinculin-binding site in the cytoskeletal protein α actinin. Biochem J 301:225–233

Menkel AR, Kroemker M, Bubeck P, Ronsiek M, Nikolai G, Jockusch BM (1994) Characterization of an F-actin-binding domain in the cytoskeletal protein vinculin. J Cell Biol 126:1231–1240

Milam LM (1985) Electron microscopy of rotary shadowed vinculin and vinculin complexes. J Mol Biol 184:543–545
Molony L, Burridge K (1985) Molecular shape and self-association of vinculin and metavinculin. J Cell Biochem 29:31–36
Nachmias VT, Golla R (1991) Vinculin in relation to stress fibers in spread platelets. Cell Motil Cytoskeleton 20:190–202
Nagafuchi A, Takeichi M, Tsukita S (1991) The 102 kD cadherin-associated protein: similarity to vinculin and posttranscriptional regulation of expression. Cell 65:849–857
Pantaloni D, Carlier MF (1993) How profilin promotes actin filament assembly in the presence of thymosin β4. Cell 75:1007–1014
Pardo JV, D'Angelo Siliciano J, Craig S (1983) Vinculin is a component of an extensive network of myofibril-sarcolemma attachment regions in cardiac muscle fibers. J Cell Biol 97:1081–108
Reinhard M, Halbrügge M, Scheer U, Wiegand C, Jockusch BM, Walter U (1992) The 46/50 kDa phosphoprotein VASP purified from human platelets is a novel protein associated with actin filaments and focal contacts. EMBO J 11:2063–2070
Rosenfeld GC, Hou DC, Kingus J, Meza I, Bryan J (1985) Isolation and partial characterization of human platelet vinculin. J Cell Biol 100:669–676
Rothkegel M, Wucherpfennig C, Kroemker M, Valenta R, Jockusch BM (1994) Association of plant profilin with the microfilament system in vertebrate cells. Eur J Cell Biol 63 (Suppl 40):98
Rozycki MD, Myslik JC, Schutt CE, Lindberg U (1994) Structural aspects of actin-binding proteins. Curr Opin Cell Biol 6:87–95
Ruhnau K, Wegner A (1988) Evidence for direct binding of vinculin to actin filaments. FEBS Lett 228:105–108
Schaller MD, Parsons JT (1993) Focal adhesion kinase: an integrin-linked protein tyrosine kinase. Trends Cell Biol 3:258–262
Schutt CE, Myslik JC, Rozycki MD, Goonesekere NCW, Lindberg U (1993) The structure of crystalline profilin-β-actin. Nature 365:810–816
Segura M, Lindberg U (1984) Separation of non-muscle isoactins in the free form or as profilactin complexes. J Biol Chem 259:3949–3954
Staiger CJ, Goodbody KC, Hussey PJ, Valenta R, Drobak BK, Lloyd CW (1993) The profilin multigene family of maize: differential expression of three isoforms. Plant J 4:631–641
Theriot JA, Mitchison TJ (1993) The three faces of profilin. Cell 75:835–838
Turner CE, Burridge K (1991) Transmembrane molecular assemblies in cell-extracellular matrix interactions. Curr Opin Cell Biol 3:849–853
Turner CE, Glenney JR, Burridge K (1990) Paxillin: a new vinculin-binding protein present in focal adhesion. J Cell Biol 111:1059–1068
Ungar F, Geiger B, Ben-Ze'ev A (1986) Cell contact- and cell shape-dependent regulation of vinculin synthesis in cultured fibroblasts. Nature 319:787–791
Van Etten RA, Jackson PK, Baltimore D, Sanders MC, Matsudaira PT, Janmey PA (1994) The COOH terminus of the c-Abl tyrosine kinase contains distinct F- and G-actin binding domains with bundling activity. J Cell Biol 124:325–340
Vinson VK, Archer SJ, Lattman EE, Pollard T, Torchia DA (1993) Three-dimensional solution structure of *Acanthamoeba* profilin I. J Cell Biol 122:1277–1283
Wachsstock DH, Wilkins JA, Lin S (1987) Specific interaction of vinculin with α actinin. Biochem Biophys Res Commun 146:554–560
Westmeyer A, Ruhnau K, Wegner A, Jockusch BM (1992) Antibody mapping of functional domains in vinculin. EMBO J 9:2071–2078
Wilkins JA, Risinger MA, Coffey E, Lin S (1987) Purification of a vinculin binding protein from smooth muscle. J Cell Biol 104:130a
Wood CK, Turner CE, Jackson P, Critchley DR (1994) Characterisation of the paxillin-binding site and the C-terminal focal adhesion targeting sequence in vinculin. J Cell Sci 107:709–717

Spectrin Interchain Binding in *Drosophila* Development

D. BRANTON[1]

1 Introduction

Spectrin, a member of the family of actin cross-linking proteins that also includes dystrophin and α-actinin, is a widely distributed protein in eukaryotes (Dubreuil 1991; Matsudaira 1991). Spectrin is generally thought to generate networks that support the plasma membrane and sustain interactions between cellular structures responsible for cell motility and shape, but this hypothesis concerning its function is derived largely from studies of the human erythrocyte. While it is clear that the membrane skeleton of erythrocytes is a spectrin network that contributes to cell shape (Branton et al. 1981; Elgsaeter et al. 1986; Marchesi 1985), cell shape in most nonerythroid cells is governed by an extensive transcellular cytoskeleton. Whether or not spectrin's network-forming capacity is the basis for in vivo function in these nonerythroid systems has not been addressed.

The formation and integrity of a spectrin network depend on inter- and intramolecular associations at two key points in the molecule: at spectrin's "tail" end, between spectrin and short actin oligomers, and at its "head" end, between spectrin and itself (Branton et al. 1981; Marchesi 1985; Byers and Branton 1985). Using a nonerythroid system, *Drosophila*, we have used in vitro assays to define the molecular basis for spectrin's interchain associations at the "head" and "tail" end (see Fig. 1 for nomenclature) of the molecule (Deng et al. 1994; Viel and Branton 1994). Transformed *Drosophila* that produce mutated spectrin were then used to show that, in vivo, spectrin's ability to form a network is as critical for its function in nonerythroid systems as it is in the erythrocyte (Deng et al. 1994).

2 Tail End Binding

Because spectrin is a large structural protein made up of repetitive and nonrepetitive segments (Marchesi 1985), it is convenient to designate these segments using a numerical nomenclature (Fig. 1). Studies that map the sequences responsible for head and tail end interchain binding have used segments of spectrin synthesized by expressing the protein encoded in portions of its cDNA. To assure correct folding, the boundaries of the synthesized segments were selected on the basis of analyses

[1] Department of Molecular and Cellular Biology, The Biological Laboratories, Harvard University, Cambridge, MA, 02138, USA.

45. Colloquium Mosbach 1994
The Cytoskeleton

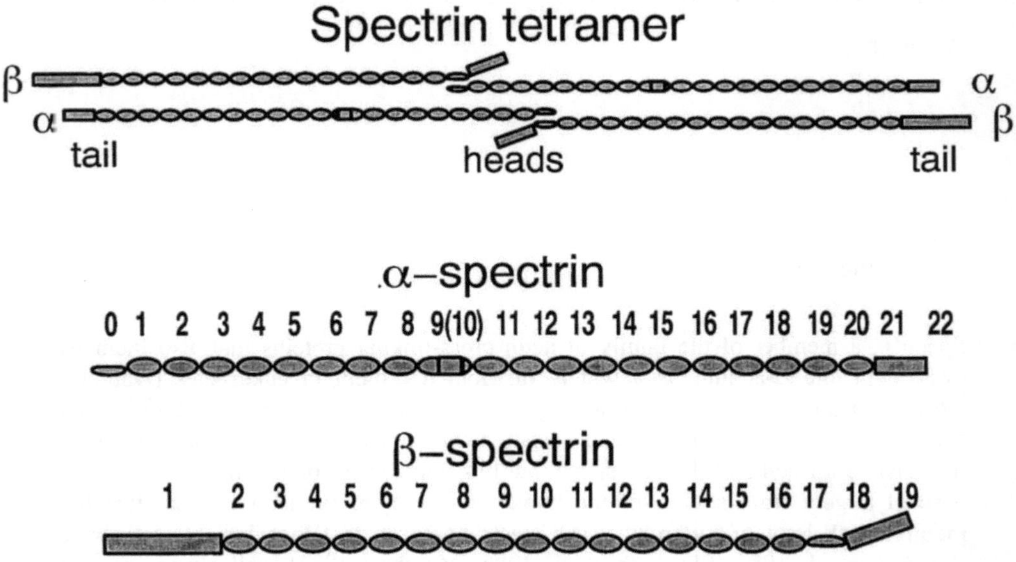

Fig. 1. Spectrin and its segment nomenclature. *Top* A cartoon showing a spectrin tetramere composed of two spectrin heterodimers (each heterodimer containing one α- and one β-spectrin chain) joined head-to-head by associations between segment 0 of α-spectrin and segment 18 of β-spectrin. *Bottom* The two spectrin chains are displayed with both their nonrepetitive segments (*rectangles*) and repetitive segments (*ovals*) designated numerically. The amino termini reside in α0 and β1, the carboxy termini in α22 and β19. Interchain binding involves segments α0 and β18 at the head end (cartooned at *top* in tetramers and shown in Fig. 4 in heterodimers) and segments α20–a22 and β1–3 at the tail end (not shown)

(Winograd et al. 1991) that defined spectrin's conformational units. The atomic structure in each of these conformational units, or segments, consists of a three helix bundle (Fig. 2) (Yan et al. 1993).

The results of several studies (Tse et al. 1990; Kotula et al. 1993; Deng et al. 1994) have shown that the head end association of the two spectrin chains is mediated by segment 18 of β-spectrin and segment 0 of α-spectrin (Figs. 1, 3). Because segment 18 of β-spectrin (β18) is homologous to two of the three helices (the A and B helices) contained in most of the other repetitive segments found in spectrins, and because segment 0 of α-spectrin (α0) is homologous to the third (C) helix, we, and others (Tse et al. 1990; Kotula et al. 1993), assume that the head end association produces a structure similar to a complete three helix segment (Fig. 3). Thus, the head end associations between α- and β-spectrin that produce tetramers appear to conserve the packing arrangements seen in each repetitive segment of spectrin (Yan et al. 1993). Furthermore, the head end interchain binding sites of *Drosophila* spectrin associate with an avidity and temperature dependence similar to those of erythroid spectrin and do, in fact, compete with the erythroid spectrin interchain binding sites in vitro (Fig. 4).

Information about the atomic structure of spectrin (Yan et al. 1993) facilitated the design of point mutations intended to disrupt interchain binding. For example, amino

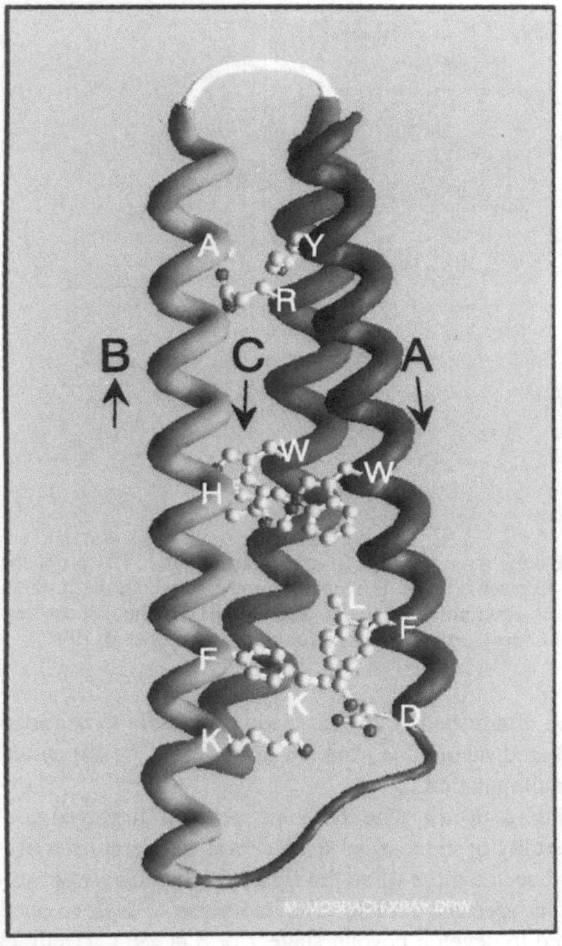

Fig. 2. Side view of one spectrin segment. Some of the side chains that pack to maintain the spacing between the α-helices are shown. Amino acids are indicated by their *single letter code*. The BC loop (*white*) has been inserted by model building. (Yan et al. 1993)

acid substitutions equivalent to erythroid spectrin mutations that are responsible for human hemolytic anemias also diminished *Drosophila* spectrin head end interchain binding in vitro (Deng et al. 1994). To test the consequence of deficient head end interchain binding in vivo, constructs expressing wild-type or head end interchain binding mutant α-spectrin were transformed into the *Drosophila* genome and tested for their ability to rescue a lethal a-spectrin null mutation. An α-spectrin minigene lacking the codons for head end interchain binding failed to rescue the first instar lethality of the null mutant, whereas a minigene with a point mutation in these codons overcame the first instar lethality of the null mutant in a temperature-dependent manner. Because spectrin's ability to form a network depends on the association between head end interchain binding sites on α- and β-spectrin, it was not surprising that transformants bearing α-spectrin minigene constructs lacking the head end binding region were unable to rescue α-spectrin null mutants. On the other hand, the use of transformants bearing an alpha chain whose association with β-spectrin was strongly

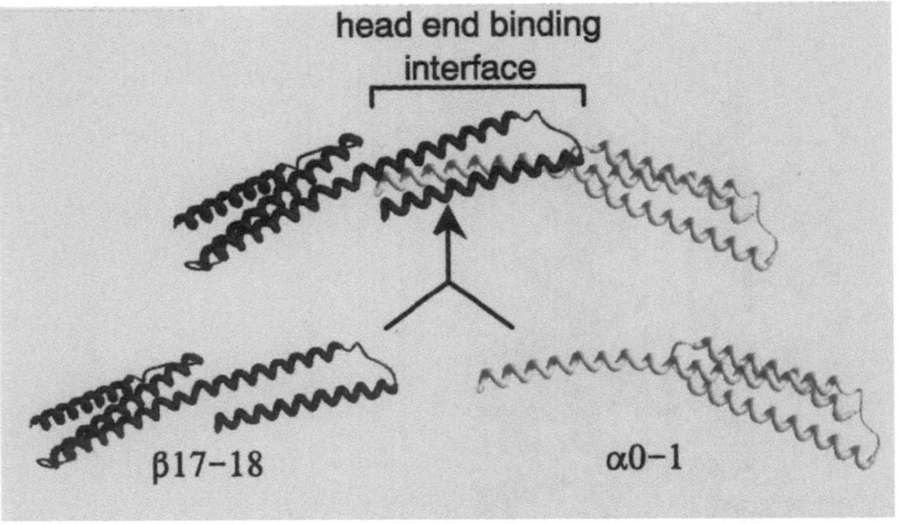

Fig. 3. Cartoon showing how β18 and α0 may give rise to interchain binding. The proposed structure of a tetrameric junction is represented by a three helix bundle to which the β chain contributes two helices and the α chain contributes one helix. Segment β19 at the carboxy terminus of β-spectrin and the first 14 residues of α0 are omitted for clarity. (Deng et al. 1994)

temperature-dependent (and much diminished at 29 °C), made it possible to examine the requirements for a spectrin-based membrane skeleton at stages of development previously inaccessible with the null mutation.

The rescued progeny expressing α-spectrin with the point mutation displayed normal gross morphology, but the fertility of the rescued females was temperature sensitive. The basis for this infertility became clear when the fly's egg chambers were examined. In flies bearing the mutant spectrin, follicle cells and nurse cells developed normally at 19 or 25 °C, but failed to develop beyond stage 7 or 8 at 29 °C (Deng et al. 1994). The arrested cell development, and eventual necrosis, occurred at a critical point during oogenesis when yolk formation in the oocyte is known to require an increase in total egg chamber volume (Mahowald and Kambysellis 1980). That these disruptions began in a localized manner, where the follicle cells and cystocytes abut, suggested that an interaction between the two cell types was disrupted, possibly because an intact spectrin network is required to sustain the convoluted plasma membrane projections that normally interlock follicle cells and cystocytes (Mahowald 1972).

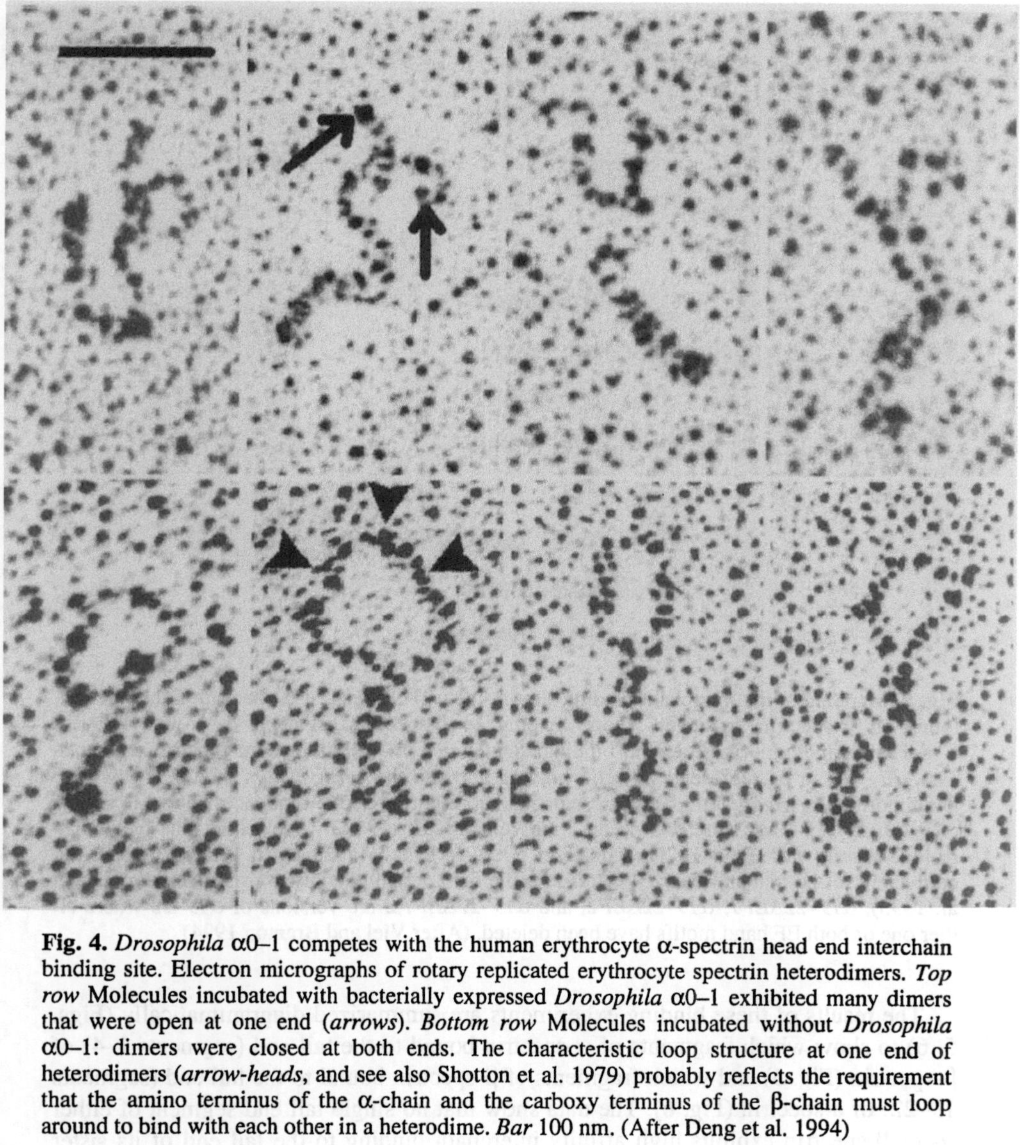

Fig. 4. *Drosophila* α0–1 competes with the human erythrocyte α-spectrin head end interchain binding site. Electron micrographs of rotary replicated erythrocyte spectrin heterodimers. *Top row* Molecules incubated with bacterially expressed *Drosophila* α0–1 exhibited many dimers that were open at one end (*arrows*). *Bottom row* Molecules incubated without *Drosophila* α0–1: dimers were closed at both ends. The characteristic loop structure at one end of heterodimers (*arrow-heads*, and see also Shotton et al. 1979) probably reflects the requirement that the amino terminus of the α-chain and the carboxy terminus of the β-chain must loop around to bind with each other in a heterodime. *Bar* 100 nm. (After Deng et al. 1994)

3 Head End Interchain Binding

Similar analyses, including immunoprecipitation assays using protein fragments synthesized in vitro from recombinant DNA, were used to map the sequences of spectrin that are required for tail end association between α- and β-spectrin (Viel and Branton 1994).

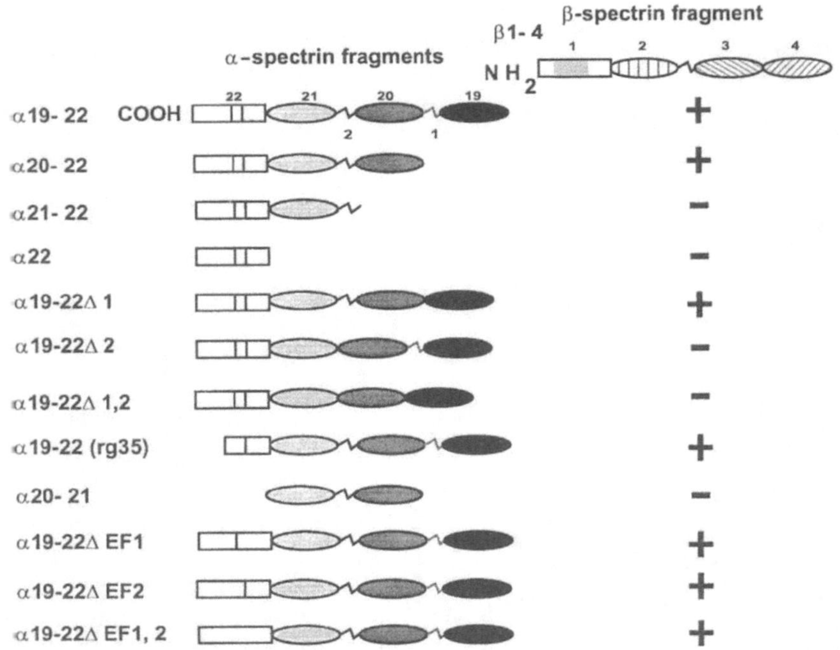

Fig. 5. Mapping of the interchain binding sites at the carboxy-end of α-spectrin. Each line cartoons an α-spectrin fragment whose binding to β-spectrin segments 1–4 was assayed. + indicates binding; – indicates no binding. *Rectangles* or *ovals* represent spectric segments. The *gray box* in β1 represents an actin binding domain; the *vertical bars* in α22 are two EF hand motifs; the *wiggly lines* between segments represent 8-residue sequences shown in Fig. 7. α*19–22Δ1*, α*19–22Δ2*, and α*19–22Δ1,2* are versions of α*19–22* where either one or both 8-residue inserts were deleted. The construct α*19–22(rg35)* is a truncated version of α*19–22*, equivalent to the mutant *l(3)dre3*^{rg35} responsible for a lethal phenotype in *Drosophila* (Lee et al. 1993). α*19–22ΔEF1*, α*19–22ΔEF2*, and α*19–22ΔEF1,2* are versions of α*19–22* where either one or both EF hand motifs have been deleted. (After Viel and Branton 1994)

The results of these binding experiments are summarized diagrammatically (Figs. 5, 6) to show which fragments of α-spectrin bound to the tail end (segments 1–4) of β-spectrin (Fig. 5) and which fragments of β-spectrin bound to the tail end (segments 19–22) of a-spectrin (Fig. 6). The data show that no single tail end segment of either α- or β-spectrin exhibits high affinity interchain binding to the tail end of its sister strand. Rather, several regions in each spectrin subunit contribute to tail end interchain binding. These regions of the spectrin molecule, including α20, α21, and part of α22 and β2, β3 and part of β1, contain unique sequences that differ from those found in the repetitive segments that make up most of the length of the spectrin molecule.

When compared with the other repeating segments of spectrin, segments α20 and α21 and β2 and β3 are longer (Fig. 7 A). If the spectrin sequences (Dubreuil et al. 1989; Byers et al. 1992) at the tail end of the molecule are aligned and displayed in accord with the boundaries of the conformational units (Winograd et al. 1991), it be-

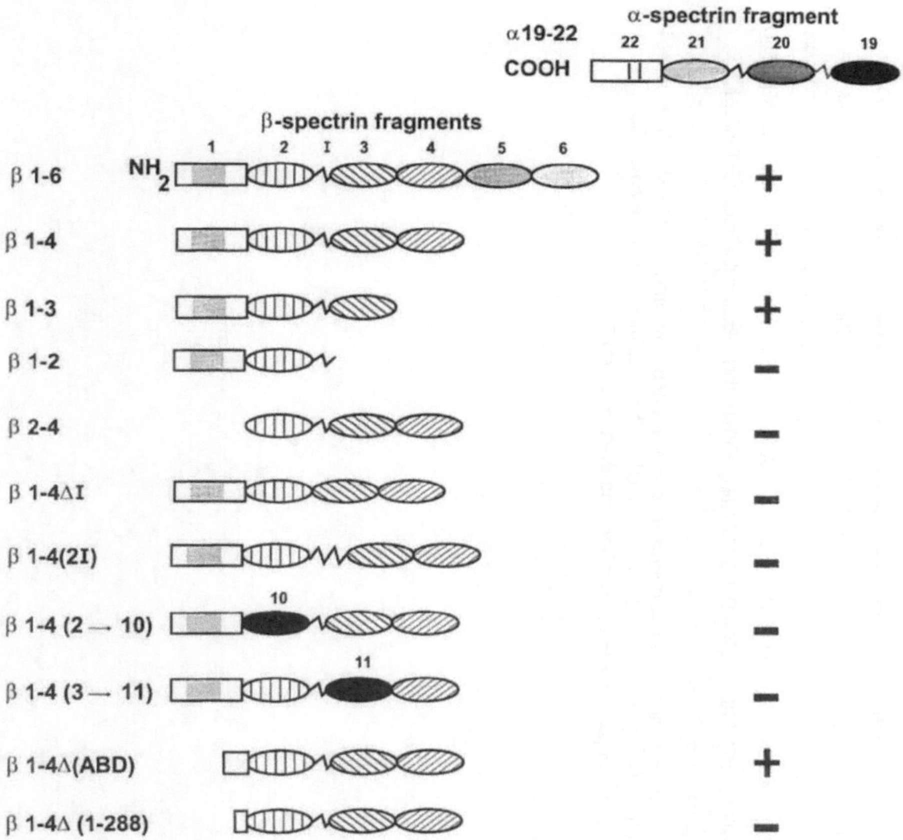

Fig. 6. Mapping of the interchain binding sites at the amino end of β-spectrin. Each line cartoons a β-spectrin fragment whose binding to α-spectrin segments 19–22 was assayed. *Rectangles, ovals, gray box, vertical bars,* and *wiggly lines* as in Fig. 5. β*1–4ΔI* is a version of β*1–4* lacking the 8-residue insert. β*1–4(2I)* is a version of β*1–4* containing two tandem repeats of the 8-residue insert. β*1–4ΔABD* is a truncated version of β*1–4* lacking the actin binding domain. β*1–4Δ288* is a truncated version of β*1–4* lacking most of the segment 1. β*1–4 (2 → 10)* and β*1–4 (3 → 11)* are versions of β*1–4* where segments 2 and 3 were replaced by segments 10 and 11 respectively. (Viel and Branton 1994)

comes evident that the greater apparent size of the repetitive segments near the amino terminus of β-spectrin and the carboxy terminus of α-spectrin is largely accounted for by sets of 8 amino acid inserts (Fig. 7, bold characters) within β-spectrin segments 2–3 and within α-spectrin segments 19–20 and 20–21. With the exception of the insert located within α19–α20, these inserts were found to be essential for tail end interchain binding (Figs. 5, 6) (Viel and Branton 1994).

The regions within α22 and β1 that are involved in interchain binding are located upstream of the first EF hand motif in α22 and downstream, immediately following the actin-binding domain, in β-spectrin. While none of the well-described structural

```
                  |------------A-------------|         |-----------------B-----------------|         |--------------C--------------|
α14 (1392-1497)           NLDLQLYMRDCELAESWMSAREAFLNADDDANAGGNVEALIKKHEDFDKA.....INGHEQKIAALQTVADQLIAQNHYA.SNLVDEKRKQVLERWRHLKEGLIEKRSRLGD
α20 (2030-2143)   MQEQFRQIEELYLTFAKKASAFNSWFENAEEDLTDPVRCNSIEEIRALRDAHAQFQAS.....LSSAEADFKALAALDQKIKSFNVGP.NPYTWFTMEALEETWRNLQKIIEERDGELAK
α21 (2144-2258)   EAKRQEENDKLRKEFAKHANLFHQWLTETRTSMMEGSGSLEQQ.LEALRVKATEVRAR....RVDLKKIEELGALLEEHLILD......NRYTEHSTVGLAQQWDQLDQLSMRMQHNLEQQIQARN
ß2 (303-416)                  YENFTSDLLKWIETTIQSLGEREFENSLAGVQGQLAQFSNYRTIEKPPKFVEKGNLEVLLFTLQSKMRANNQKPYTPKEGKMISDINKAWERLEKAEHERELALREELIRQEKL
ß3 (417-522)              EQLAARFDRKASMRETWLSENQRLVSQDNFGFDLAAVEAAAKKHEAIETD.....IFAYEERVQAVVAVCDELESERYHD.VKRILLRKDNVMRLWTYLLELLRARRMRLEI
```

Fig. 7. Alignment of *Drosophila* α14 sequence with *Drosophila* sequences from segments involved in interchain binding. The structure of α14 is known (Yan et al. 1993), and the boundaries of its three α-helices (*A, B,* and *C*) are shown above α14. The 8-residue inserts are *bold*; those involved in interchain binding are *bold and underlined*. Residue numbers are *in parentheses*. (After Viel and Branton 1994)

features within these non-repetitive segments (the EF hand motifs in α segment 22, and the actin binding domain in β segment 1) are directly required for interchain binding, the interaction between segments β1 and α22 that brings into apposition the actin binding site on β-spectrin and the EF hand motifs on α-spectrin may contribute to the calcium regulation of actin binding to spectrin (Fowler and Taylor 1980; Fishkind et al. 1987) or to the homologous features of α-actinin (Witke et al. 1993). Furthermore, the diminished interchain binding observed when the first EF hand was deleted (Fig. 5) suggests that this domain may have a stabilizing effect on the interchain binding site in segment α22.

The mapping of interchain binding sites that emerges from these studies is currently being used in genetic experiments to test in vivo function in *Drosophila*.

4 Conclusions

The observations that oogenesis fails to progress beyond a specific stage of development suggests that the functional requirements on spectrin head end binding varies in a stage specific manner. Furthermore, other observations, not reviewed here, imply that the functional demand for spectrin may vary from one tissue type to another. Because spectrin is a large structural molecule that normally exhibits many different specific intra- and intermolecular associations in addition to head end and tail end binding, the question "Why is spectrin essential for development?" cannot have one single answer. It will be more instructive to ask "Which of spectrin's several binding functions are critical for normal development, and why?" This question can now be addressed by the combination of in vitro experiments that make it possible to map specific binding sites, and in vivo assessments that use genetic approaches to determine how phenotype is affected by defects or alterations at specific sites within the spectrin molecule.

References

Branton D, Cohen CM & Tyler J (1981) Interaction of cytoskeletal proteins on the human erythrocyte membrane. Cell 24:24–32

Byers TJ & Branton D (1985) Visualization of the protein associations in the erythrocyte membrane skeleton. Proc Natl Acad Sci USA 82:6153–6157

Byers TJ, Brandin E, Lue RA, Winograd E & Branton D (1992) The complete sequence of *Drosophila* beta-spectrin reveals supra-motifs comprising eight 106-residue segments. Proc Natl Acad Sci USA 89:6187–6191

Deng H, Lee JK, Goldstein LSB & Branton D (1995) Structure and function of *Drosophila* spectrin head-end interchain binding sites. J Cell Biol 128:71–80

Dubreuil RR, Byers TJ, Sillman AL, Bar-Zvi D, Goldstein LSB & Branton D (1989) The complete sequence of *Drosophila* alpha-spectrin: conservation of structural domains between alpha-spectrins and alpha-actinin. J Cell Biol 109:2197–2205

Dubreuil RR (1991) Structure and evolution of the actin crosslinking proteins. Bio Essays 13:219–226

Elgsaeter A, Stokke BT, Mikkelsen A & Branton D (1986) The molecular basis of erythrocyte shape. Science 234:1217–1223

Fishkind DJ, Bonder EM & Begg DA (1987) Isolation and characterization of sea urchin egg spectrin: calcium modulation of the spectrin-actin interaction. Cell Motil Cytoskeleton 7:304–314

Fowler V & Taylor DL (1980) Spectrin plus band 4.1 cross-link actin. Regulation by micromolar calcium. J Cell Biol 85:361–376

Kotula L, DeSilva TM, Speicher DW & Curtis PJ (1993) Functional characterization of recombinant human red cell alpha-spectrin polypeptides containing the tetramer binding site. J Biol Chem 268:14788–14793

Lee JK, Coyne RS, Dubreuil RR, Goldstein LSB & Branton D (1993) Cell shape and interaction defects in alpha-spectrin mutants of *Drosophila melanogaster.* J Cell Biol 123:1797–1809

Mahowald AP (1972) Ultrastructural observations on oogenesis in *Drosophila.* J Morphol 137:29–48

Mahowald AP & Kambysellis MP (1980) Oogenesis. pp 142–224. In: Ashburner M & Wright TRF (eds) Genetics and biology of *Drosophila.* Academic Press, London

Marchesi VT (1985) Stabilizing infrastructure of cell membranes. Ann Rev Cell Biol 1:531–561

Matsudaira P (1991) Modular organization of actin crosslinking proteins. Trends Biochem Sci 16:87–92

Shotton DM, Burke B & Branton D (1979) The molecular structure of human erythrocyte spectrin: biophysical and electron microscopic studies. J Mol Biol 131:303–329

Tse WT, Lecomte MC, Costa FF, Garbarz M, Feo C, Boivin P, Dhermy D & Forget BG (1990) Point mutation in the beta-spectrin gene associated with alpha-I/74 hereditary elliptocytosis – implications for the mechanism of spectrin dimer self-association. J Clin Invest 86:909–916

Viel A & Branton D (1994) Interchain binding at the tail end of the *Drosophila* spectrin molecule. Proc Natl Acad Sci USA 9:10839–10843

Winograd E, Hume D & Branton D (1991) Phasing the conformational unit of spectrin. Proc Natl Acad Sci USA 88:10788–10791

Witke W, Hofmann A, Koppel B, Schleicher M & Noegel AA (1993) The Ca^{2+}-binding domains in non-muscle type alpha-actinin: biochemical and genetic analysis. J Cell Biol 121:599–606

Yan Y, Winograd E, Viel A, Cronin T, Harrison SC & Branton D (1993) Crystal structure of the repetitive segments of spectrin. Science 262:2027–2030

Intermediate Filament Proteins: Structure, Function, and Evolution

K. WEBER[1]

1 Introduction

Intermediate filaments (IF) are the third major component of the cytoskeleton. Cytoplasmic IF are present in nearly all vertebrate cells and the molar amount of these proteins can often surpass that of actin and tubulin. IF have a diameter around 10 to 11 nm, a value intermediate between the diameters of F-actin and microtubules. In the past 2 years major progress has been made in understanding the structure of IF, while the characterization of human epidermal keratin mutants has opened a direct way to understand IF function. In addition, analysis of IF proteins from invertebrates has yielded the first clues to IF evolution.

2 The IF Multigene Family of Vertebrates

Table 1 shows that the multigene family of IF proteins currently covers close to 50 distinct members in a mammal. Biochemical properties, protein sequences, expression rules, and intron patterns of the genes argue for a least five subtypes. Types I and II, which encode the epithelial keratins I and II respectively, are particularly large. Keratin filaments are obligatory heteropolymers in which the heterodimer, formed by one type I and one type II polypeptide, is the essential building block. There are at least four type III proteins. They can form in vivo and in vitro homopolymeric IF. Examples are the desmin of various muscle cells, the GFAP of glial cells, the peripherin from the peripheral nervous system, and the mesenchymal vimentin. Type IV covers the neurofilament triplet proteins NF-L, NF-M, and NF-H as well as α-internexin and nestin. Two cytoplasmic IF proteins restricted to the eye lens – filensin and phakinin – are not yet classified. Finally, the lamins are covered as nuclear IF proteins by type V. The two B-type lamins, B_1 and B_2, are essentially constitutively expressed. The larger A-lamin, which has an extra domain, seems restricted to differentiated cells. Lamin C, a truncated form of lamin A, arises by differential RNA processing. It has only been observed in mammals. Two further B-type lamins seem unique. The major lamin of the oocyte and early embryo of *Xenopus* is lamin B_3, while mouse spermatocytes contain a lamin which arises by differential splicing from the B_2 gene (reviewed by Fuchs and Weber 1994).

---

[1] Max Planck Institute for Biophysical Chemistry, Department of Biochemistry, P.O. Box 2841, D-37018 Goettingen, FRG.

45. Colloquium Mosbach 1994
The Cytoskeleton

Table 1. The IF multigene family in vertebrates

I	Epithelial keratins K9–K20 plus	four type I hair keratins
II	Epithelial keratins K1–K8 plus	four type II hair keratins
III	Vimentin "mesenchymal" GFAP glia	desmin myogenic peripherin peripheral nervous system
IV	Neurofilament proteins triplet proteins NF-L, NF-M, NF-H α-internexin nestin	 adult neurons embryonic brain neuroepithelial cells
V	Nuclear lamins B1, B2 A/C special B2 in spermatocytes,	 constitutively expressed differentiated cells special B3 in oocyte

Not classified: filensin plus 47 kDa/protein (phakinin) in eye lens.

Nearly all lamins end with a carboxy terminal CAAX box which is subject to three posttranslational modifications (isoprenylation of the cysteine, removal of the terminal three amino acids, and O-methylation of the terminal cysteine). The resulting hydrophobic cysteine derivative is either directly or indirectly responsible for membrane anchorage and lamina incorporation. Targeting of the lamins into the nuclear membrane requires the presence of the nuclear location signal and the CAAX box (for recent reviews on lamins see McKeon 1991; Nigg 1992; Nigg, this Vol). Cytoplasmic IF proteins lack both topogenic sequences.

The three domain structures of IF protein is defined by proteolytic dissection and sequence principles. The α-helical rod domain is flanked by the nonhelical head and tail domain. The rod domain can form a double-stranded segmented coiled coil based on helices 1A, 1B, 2A, and 2B, which are separated by short spacers. The helices have heptad repeats in which the *a* and *d* positions are usually occupied by hydrophobic residues. These provide the major contacts between the two helices, which are arranged in parallel and in register. In the various rod domains of type I to IV proteins, sequence principles rather than actual sequences are conserved. True consensus sequences are particularly obvious at the two ends of the rod domains. The length of the rod domain is usually 310 residues in the cytoplasmic IF proteins of vertebrates. Nuclear lamins and the cytoplasmic IF proteins of invertebrates (see below) have an additional 42 residues (6 heptads) in the coil 1B domain.

The N-terminal head- and the C-terminal tail domains can display extreme variability both in sequence and in length. Heads can vary from 10 to 170 residues and tails from 10 to 1500 residues. These peripheral domains can provide for an individuality of the environment of different IF. Extreme examples are the epidermal keratins with their repeating uncharged glycine-rich loops in head- and tail domains and the cysteine-rich terminal domains of wool keratins, which seem to form disulfides with special matrix proteins. In contrast, the neurofilament tails are highly charged, and NF-H in particular contains additional multiple KSP repeats which are phosphorylated. While the IF structure relies primarily on a special packing of dimers, which give rise to different tetramers (see below), the terminal domains also make a contribution. Proteolytic experiments show that the head domain of type III proteins is required for elongation beyond the tetramer stage and thus for IF formation. The tail domains seem to control the filament diameter. For a detailed presentation of IF and original references see the review by Fuchs and Weber (1994).

3 IF Structure

A satisfactory model of IF structure has emerged from chemical crosslinking studies and electronmicroscopical results. The model is based on three tetramers, which describe lateral contacts between antiparallel oriented dimers. Two tetrameric particles are staggered by about half a unit length and rely either on coil 1 interactions or on coil 2 interactions. The shorter third particle is unstaggered and is based on coil 1/coil 2 interactions. A fourth tetrameric particle describing a longitudinal interaction is based on a small N/C overlap and shows the contact between the two consensus sequences at the end of the rod domain. It reflects the early assembly stage of lamins by a head to tail type assembly. From these four tetramers, one can construct either a surface lattice (Steinert et al. 1993) or a filament model emphasizing the presence of protofibrils and protofilaments (Heins and Aebi 1994).

4 IF Function

For many years the cellular function of IF has been a matter for debate. This situation has changed with the discovery that the hereditary human blistering diseases of the skin arise from point mutations in specific epidermal keratins. Due to the contributions of several laboratories, point mutations have been characterized in the following keratins: K1, K2, K5, K9, K10, and K14. These epidermal skin diseases are covered by Lane in this Volume. The position of the natural human point mutations is nonrandom. A sizeable fraction occurs in the two consensus-type sequences at the ends of the rod domain. Thus it seems that severe mutations lie in the multiple contact zones predicted by the current IF structural model.

Recent studies show that neurofilaments are required for proper radial growth of axons. Overexpression of neurofilament molecules in transgenic mice leads to abnormal accumulation of neurofilaments and can cause motor neurone dysfunction (reviewed by Lee and Cleveland 1994; Cleveland, Lee, Marszalek and Xu, this Vol.). Thus, several interesting phenotypes and diseases are connected with defects in specific IF molecules.

5 IF from invertebrates; some aspects of IF evolution

Currently, molecular knowledge of cytoplasmic IF proteins extends from vertebrates only to some invertebrate species. Selected electron micrographs indicate, however, that IF may be more widely spread in eukaryotes. Direct protein sequence analysis of IF proteins from the esophagus of the mollusk *Helix promatia* and the muscle tissue of the nematode *Ascaris lumbricoides* documented two unexpected properties (Weber et al. 1988; Weber et al. 1989). These cytoplasmic IF proteins display some lamin-like features, which are clearly absent from all cytoplasmic IF proteins of vertebrates. They have an extra six heptads in the coil 1b domain and the lamin-like structure extends to the tail domains. Here a lamin homology segment of some 120 residues corresponds to the sequence of B-type lamins located between the nuclear location signal and the CAAX box. Over this region, *Drosophila* lamin Dmo shows nearly 40% sequence identity with various vertebrate lamins. The nematode IF protein B displays 30 to 40% identity when compared with *Drosophila* or vertebrate lamins. Thus it seemed that the archetype cytoplasmic IF protein arose early in eukaryotic evolution from a mutated lamin, which had lost the two topogenic sequences related with lamin functionality (nuclear location signal and CAAX box). This view was reinforced when the organizations of the *Helix aspersa* IF1 gene and the lamin B3 gene of *Xenopus* were established (Dodemont et al. 1990; Döring and Stick 1990). Eight of the ten introns present in both genes are identically placed. Intron 7 of the *Helix* gene is unique and occurs at the position where the lamin gene encodes the nuclear location signal. The last intron of the lamin gene is also unique. It borders the mini-exon encoding the CAAX motif. Thus it was suggested that the acquisition of a new splice site freed the archetype cytoplasmic IF protein of the nuclear location signal, while the introduction of an earlier stop codon removed the CAAX box. The resulting IF protein stayed in the cytoplasm and formed IF without unwanted membrane interactions. This view is consistent with the behavior of certain lamin mutants studied for different reasons in transfection assays. The mutant lamins form filament bundles in the cytoplasm (McKeon 1991).

The list of invertebrate IF proteins has meanwhile increased by a cDNA sequence for IF1 from *Aplysia californica* (Riemer et al. 1991) and the eye lens-specific IF proteins of the octopus and squid (Tomarev et al. 1993). In addition, it was shown that the three neurofilament proteins of the squid are encoded by a single gene, which is nearly completely characterized (Szaro et al. 1991; Way et al. 1992). Finally we have described 8 IF genes in the nematode *C. elegans*. They give rise to a total of 12 IF proteins (Dodemont et al. 1994). All these IF proteins from mollusks and nematodes show the long coil 1b version. The extra number of residues is 42 except for the *C. elegans* protein b_2, which shows only 39 residues (Dodemont et al. 1994). While the long coil 1b version is well conserved, some of the IF proteins have already lost the lamin homology segment in the tail domain and acquired entirely new sequences. This is the case in two of the three neurofilament proteins of the squid (Szaro et al. 1991; Way et al. 1992) and in 3 of the 12 *C. elegans* IF proteins (Dodemont et al. 1994). Thus, it seems easier to lose the tail homology segment that to shorten the coil 1b domain.

Nematodes and mollusks are phyla of the protostomic branch of metazoa. We still lack molecular information on cytoplasmic IF proteins from invertebrate species of the deuterostomic branch, which leads to the vertebrates. Somewhere on this branch, the coil 1b domain was shorted by 42 residues and subsequent gene duplication events gave rise to the predecessors of type I to IV IF genes. We have recently characterized an IF gene from *Branchiostoma lanceolatum*, a member of the cephalochordates, which is a sister group of the vertebrates. This IF gene of an early chordate already shows the short coil 1b domain (Riemer et al. 1992). Thus the deletion most likely occurred prior to the emergence of the chordates, possibly already with the origins of deuterostomic metazoa.

One of the drawbacks in trying to understand the evolution of IF proteins and their complexity is the lack of data on such proteins in other eukaryotic kingdoms such as protists, plants, and fungi. Only when sequence information on such proteins becomes available can we assess the current speculations of IF evolution. It seems particularly important to decide whether the longer coil 1b version of protostomic IF proteins really reflects evolutionary lamin descent as discussed above, or is due to an insertion which occurred selectively at the root of the protostomic branch.

To complete this chapter, we note that molecular knowledge about nuclear lamins is also very restricted. Outside the vertebrates only three lamins are known; all of them come from invertebrates. Two lamin genes are known for *Drosophila*: lamin Dmo (Osman et al. 1990) and lamin C (Bossie and Sanders 1993; Riemer and Weber 1994). One lamin gene has been established in *C. elegans* (Riemer et al. 1993).

References

Bossie CA, Sanders MM (1993) A cDNA from *Drosophila* melanogaster encodes a lamin C-like intermediate filament protein. J Cell Sci 104:1263–1272

Dodemont H, Riemer D, Ledger N, Weber K (1994) Eight genes and alternative RNA processing pathways generate an unexpected large diversity of cytoplasmic intermediate filament (IF) proteins in the nematode *Caenorhabditis elegans*. EMBO J 13:2625–2638

Dodemont H, Riemer D, Weber K (1990) Structure of an invertebrate gene encoding cytoplasmic intermediate filament (IF) proteins: implications for the origin and the diversification of IF proteins. EMBO J 9:4083–4094

Döring V, Stick R (1990) Gene structure of nuclear lamin LIII of *Xenopus laevis*; a model for the evolution of IF proteins from a lamin-like ancestor. EMBO J 9:4073–4081

Fuchs E, Weber K (1994) Intermediate filaments: structure, dynamics, function and disease. Ann Rev Biochem 63:345–382

Heins S, Aebi U (1994) Making heads and tails of intermediate filament assembly dynamics and networks. Curr Opin Cell Biol 6:25–33

Lee MK, Cleveland DW (1994) Neurofilament function and dysfunction: involvement in axonal growth and neuronal disease. Curr Opin Cell Biol 6:34–40

McKeon F (1991) Nuclear lamin proteins: domains required for nuclear targeting, assembly and cell-cell-regulated dynamics. Curr Opin Cell Biol 3:82–86

Nigg EA (1992) Assembly-disassembly of the nuclear lamina. Curr Opin Cell Biol 4:105–109

Osman M, Paz M, Landesman Y, Fainsod A, Gruenbaum Y (1990) Molecular analysis of the *Drosophila* nuclear lamin gene. Genomics 8:217–224

Riemer D, Dodemont H, Weber K (1990) Cloning of the non-neuronal intermediate filament protein of the gastropod *Aplysia californica*; identification of an amino acid residue essential for the IFA epitope. Eur J Cell Biol 56:351–357

Riemer D, Dodemont H, Weber K (1992) Analysis of the cDNA and gene encoding a cytoplasmic intermediate filament (IF) protein from the cephalochordate *Branchiostoma lanceolatum*; implications for the evolution of the IF protein family. Eur J Cell Biol 58:128–135

Riemer D, Dodemont H, Weber K (1993) A nuclear lamin of the nematode *Caenorhabditis elegans* with unusual structural features; cDNA cloning and gene organisation. Eur J Cell Biol 62:214–223

Riemer D, Weber K (1994) The organisation of the gene for *Drosophila* lamin C: limited homology with vertebrate lamin genes and lack of homology versus the *Drosophila* lamin Dmo gene. Eur J Cell Biol 63:299–306

Steinert PM, Marekow LN, Fraser RDB, Parry DAD (1993) Keratin intermediate filament structure. Cross-linking studies yield quantitative information on molecular dimensions and mechanisms of assembly. J Mol Biol 230:436–452

Szaro BG, Pant HC, Way J, Battey J (1991) Squid low molecular weight neurofilament proteins are a novel class of neurofilament protein. J Biol Chem 266:15035–15041

Tomarev SI, Zinovieva RD, Piatigorsky J (1993) Primary structure and lens-specific expression of genes for an intermediate filament protein and a β-tubulin in cephalopods. Biochim Biophys Acta 1216:245–254

Way J, Hellmich MR, Jaffe H, Szaro B, Pant HC, Gainer H, Battey J (1992) A high molecular-weight squid neurofilament protein contains a lamin-like rod domain and a tail domain with Lys-Ser-Pro repeats. Proc Natl Acad Sci USA 89:6963–6967

Weber K, Plessman U, Dodemont H, Kossmagk-Stephan K (1988) Amino acid sequences and homopolymer-forming ability of the intermediate filament proteins from an invertebrate epithelium. EMBO J 7:2995–3001

Weber K, Plessman U, Ulrich W (1989) Cytoplasmic intermediate filament proteins of invertebrates are closer to nuclear lamins than are vertebrate intermediate filament proteins; sequence characterization of two muscle proteins of a nematode. EMBO J 8:3221–3227

The Nuclear Lamina: Regulation of Assembly by Posttranslational Modifications

E. A. NIGG and H. HENNEKES[1]

1 Introduction

The nuclear lamina is a proteinaceous meshwork underlying the inner nuclear membrane (Gerace and Burke 1988; Nigg 1989, 1992a, b; Moir and Goldman 1993). Its major constituents, the nuclear lamins, display a high degree of structural similarity to cytoplasmic intermediate filament (IF) proteins and hence are classified as type V IFs. The common feature of IF proteins is a central α-helical rod domain (of about 360 amino acids) which contains a heptad repeat of hydrophobic amino acids favoring the formation of two-stranded coiled-coil structures (Steinert et al. 1993; Stewart 1993; Heins and Aebi 1993; Fuchs and Weber 1994). When compared to vertebrate cytoplasmic IFs, the rod domain of lamins contains an extra 42 amino acids; however, an identically increased rod length is also seen in invertebrate cytoplasmic IFs (Weber et al. 1989), suggesting that lamins may be ancestral members of the IF family. This view is supported further by a striking conservation of the intron-exon structure among vertebrate lamins and invertebrate cytoplasmic IFs (Döring and Stick 1990; Dodemont et al. 1990; Lin and Worman 1993).

The central α-helical rod domain of the different IFs is flanked by N- and C-terminal regions of variable lengths. In the case of nuclear lamins, the N-terminus is rather short (20–30 amino acids) while the C-terminus is longer (some 190–270 residues) and contains some of the structural elements that are characteristic of lamins. These include a nuclear localization signal C-terminal to the rod domain (Loewinger and McKeon 1988) and a CAAX-motif (C = cysteine, A = aliphatic amino acid, X = any amino acid) at the C-terminus. This latter motif is subject to a series of posttranslational modifications, notably isoprenylation (specifically: the addition of a C15 farnesyl-group), proteolytic trimming, and carboxyl-methylation (for review see McKeon 1991; Nigg et al. 1992).

Lamins have been identified in various vertebrate and invertebrate organisms, but as yet no lamin cDNAs have been isolated from either fungi or plants (although plants are almost certain to express lamin-like proteins, e.g., Beven et al. 1991; McNulty and Saunders 1992). Lamins have been designated as A- or B-type, according to differences in primary structure and biochemical behavior (see below); a differentially spliced version of lamin A, lacking part of the C-terminus, is called lamin C (for review see McKeon 1991; Nigg 1992a, b). B-type lamins are constitutively expressed in virtually all somatic cells, and, interestingly, specialized forms of B-type lamins

---

[1] Swiss Institute for Experimental Cancer Research (ISREC), 155, Chemin des Boveresses, CH-1066 Epalinges, Switzerland.

45. Colloquium Mosbach 1994
The Cytoskeleton

have been found in oocytes and spermatocytes of *Xenopus laevis* (Smith and Benavente 1992, Vester et al. 1993; Moss et al. 1993; Furukawa and Hotta 1993). In contrast, expression of lamins A/C is usually restricted to differentiating cells (see Nigg 1989; Röber et al. 1989; Peter and Nigg 1991; Mattia et al. 1992 and references therein). The functional significance of the differential expression of lamin isoforms remains unclear. It has been speculated that this phenomenon may relate to differences in the organization of interphase chromatin in different cell types (e.g., Nigg 1989; Röber et al. 1991), but direct experimental support for this attractive hypothesis is lacking. It is noteworthy in this context that constitutive, ectopic expression of an A-type lamin in an undifferentiated embryonal carcinoma cell line (which does not normally express this isoform until it is induced to differentiate) did not affect either the differentiation state or potential of these cells (Peter and Nigg 1991). Elimination of individual lamin genes by homologous recombination may be required to clarify this important unresolved issue.

Interestingly, A- and B-type lamins differ also in their behavior during mitosis. When the nuclear envelope breaks down at the onset of mitosis, A-type lamins are solubilized as dimers and distributed throughout the cell, whereas B-type lamins remain attached to nuclear envelope-derived membrane structures (Gerace and Blobel 1980; Stick et al. 1988). As will be discussed below, both the assembly and the cell cycle-dependent dynamics of the nuclear lamina are extensively controlled by different types of posttranslational modifications, notably phosphorylation, isoprenylation, and carboxyl methylation, as well as proteolytic processing.

2 Results and Discussion

2.1 Phosphorylation of Nuclear Lamins

During interphase of the cell cycle, lamin proteins are phosphorylated on multiple sites, and some of these sites appear to be targets of protein kinase C (PKC) and/or cAMP-dependent protein kinase (PKA) (Hornbeck et al. 1988; Eggert et al. 1993; Hennekes et al. 1993; Hocevar et al. 1993; for review see Nigg 1992a). The precise function of these interphase phosphorylations remains to be fully understood, but some interesting clues have recently emerged. PKC is unable to cause significant disassembly of in vitro assembled lamin polymers (Peter et al. 1991), and does not appear to play a significant role in the disassembly of the lamina during mitosis (Peter et al. 1990). As indicated by recent work on human lamin B_1, however, it is plausible that phosphorylation of PKC sites immediately C-terminal to the central rod domain may cause loosening up of the lamina structure (Hocevar et al. 1993). Another study has shown that phosphorylation of chicken lamin B_2 on PKC-sites reduces the nuclear uptake of lamin B_2 in vitro and that PKC activation impairs the nuclear import of newly synthesized lamin proteins in vivo (Hennekes et al. 1993). This observation can be explained by the presence of PKC phosphorylation sites in the immediate proximity to the nuclear localization signal of lamin B_2 (see Fig. 1).

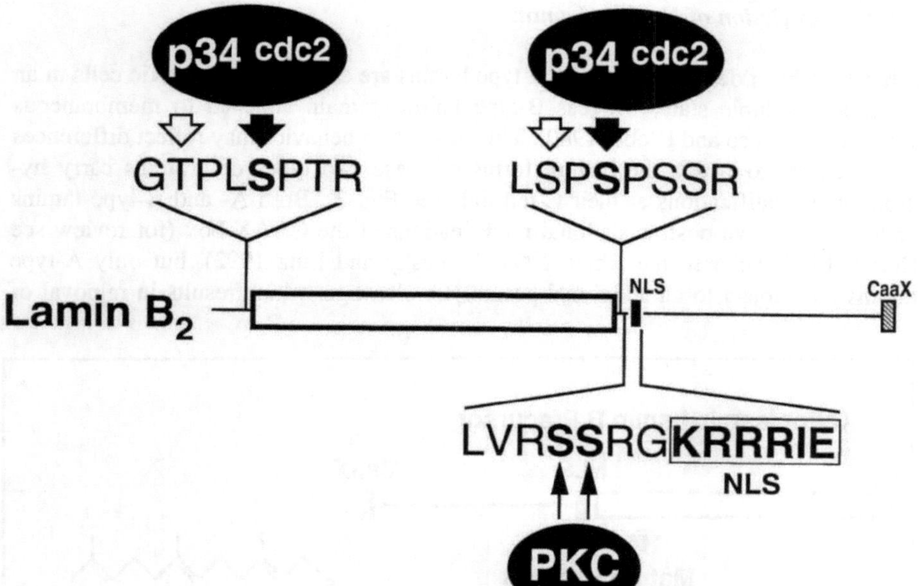

Fig. 1. Localization of major phosphorylation sites in chicken lamin B_2. Mitotic phosphorylation sites observed in vivo include serine and threonine residues that are located on either side of the central a-helical rod domain (Peter et al. 1990; Ward and Kirschner 1990). Major and minor phosphorylation sites are marked by *filled* and *open arrows*, respectively. These same residues are phosphorylated also by $p34^{cdc2}$/cyclin B in vitro, and their phosphorylation causes disassembly of lamin polymers (Peter et al. 1991). Intriguingly, the same sites may also be targeted by MAP kinases, and it is possible that MAP kinases may contribute to lamina disassembly during meiosis (Peter et al. 1992). PKC phosphorylation sites have been mapped to serine residues located in close proximity to the nuclear localization signal (NLS) of lamin B_2 (indicated by *arrows*). Phosphorylation of these sites interferes with nuclear transport of lamin B_2 in vitro as well as in vivo (Hennekes et al. 1993). (Nigg 1992a)

During mitosis, when the nuclear envelope breaks down, the phosphorylation state of lamins undergoes both qualitative and quantitative changes (Gerace and Blobel 1980; Ottaviano and Gerace 1985; Peter et al. 1990; Ward and Kirschner 1990). Work from several laboratories concurs to demonstrate that disassembly of the nuclear lamina is likely to result from the direct phosphorylation of nuclear lamins by the $p34^{cdc2}$/cyclin B kinase, a key regulator of the eukaryotic cell cycle (for review see Nigg 1991; Moir and Goldman 1993). At the end of mitosis, lamins are dephosphorylated and the lamina is rebuilt from preexisting material (Ottaviano and Gerace 1985; for review see Pfaller and Newport 1992). Phosphorylation of lamins $p34^{cdc2}$/cyclin B occurs on both sides of the central rod domain (see Fig. 1) and by causes disassembly of in vitro-formed head-to-tail polymers to dimers (Peter et al. 1991), as well as disassembly of the lamina in vivo (Peter et al. 1990; Heald and McKeon 1990).

2.2 *Isoprenylation of Nuclear Lamins*

It has long been demonstrated that A-type lamins are distributed in mitotic cells in an apparently soluble state, whereas B-type lamins remain attached to membraneous structures (Gerace and Blobel 1980). This distinctive behavior may reflect differences in the extent to which the mature forms of these two types of proteins carry hydrophobic modifications at their C-termini (see Fig. 2). Both A- and B-type lamins undergo extensive posttranslational modifications at the CAAX box (for review see Clarke 1992; Schafer and Rhine 1994; Sinensky and Lutz 1992), but only A-type lamins are subject to an additional proteolytic cleavage, which results in removal of

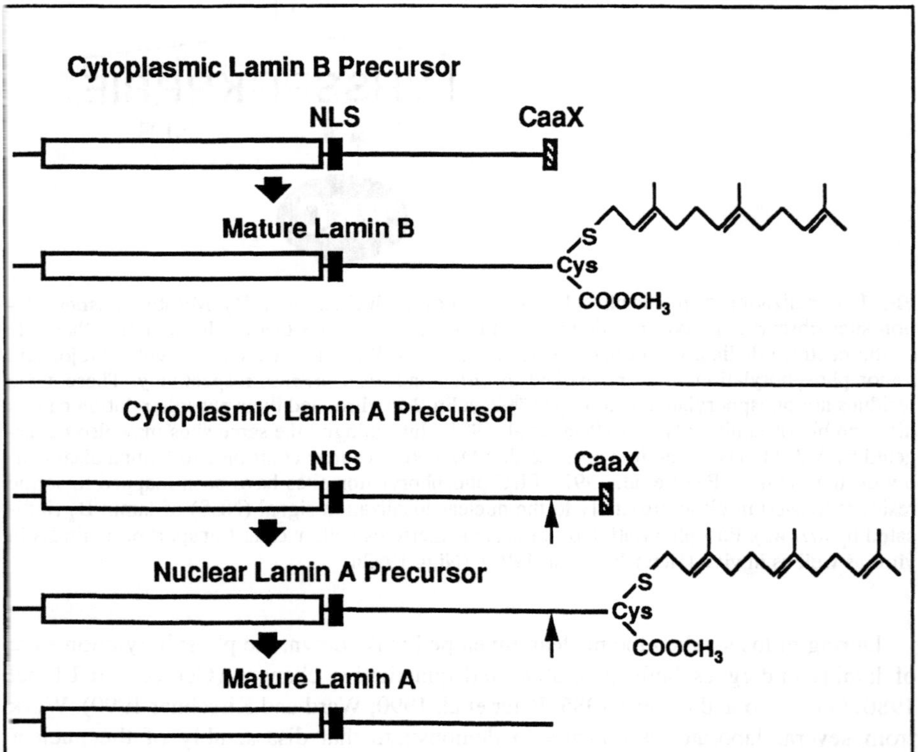

Fig. 2. Summary of the posttranslational processing events occurring at the C-termini of lamin proteins. Both A- and B-type lamins are synthesized with C-terminal CAAX boxes which are substrates for the addition of a farnesyl group to the cysteine (in a thioether linkage), the proteolytic removal of the last three amino acids (-AAX), and the carboxyl-methylation of the farnesylated cysteine. Whether these events occur in the cytoplasm and on microsomal membranes (as is the case for other, cytoplasmically located isoprenylated proteins), or whether lamins are first imported into the nucleus before they are modified, remains to be determined (for discussion see Sinensky et al. 1994). What is clear, however, is that A-type lamins undergo an additional processing event which occurs after the incorporation of A-type lamins into the nuclear lamina (Lehner et al. 1986) and consists in the proteolytic removal of the hydrophobically modified C-terminus (Vorburger et al. 1989; Weber et al. 1989). (Nigg 1992a)

the modified C-terminus (Vorburger et al. 1989, Weber et al. 1989; Hennekes and Nigg 1994; Sinensky et al. 1994). Interestingly, isoprenylation (but not proteolytic trimming or carboxyl methylation) of CAAX-box containing proteins can occur in reticulocyte lysates, i.e., a soluble cell-free system (Vorburger et al. 1989). Addition of microsomal membranes to such a system allows the subsequent proteolytic removal of the C-terminal 3 amino acids and carboxyl-methylation (Hancock et al. 1991a). It is not entirely clear where in the cell the processing of the lamin CAAX box occurs (Sinensky et al. 1994), but it is well established that these hydrophobic modifications are necessary for targeting newly synthesized lamins to the nuclear envelope (Krohne et al. 1989; Holtz et al. 1989; Kitten and Nigg 1991).

2.3 To What Extent Does Isoprenylation Contribute to the Membrane Attachment of Nuclear Lamins?

It is attractive to speculate that the persistent farnesyl-modification of B-type lamins may be responsible for their permanent attachment to the nuclear membrane. Conversely, the lack of such a hydrophobic modification might explain why mature lamin A is solubilized once mitotic phosphorylation causes disassembly of the lamina. To test this model, a point mutation was recently introduced into the lamin A-specific proteolytic cleavage site of the lamin A precursor: this mutation abolished the proteolytic removal of the hydrophobically modified C-terminus from lamin A, and concomitantly conferred membrane binding properties to the permanently farnesylated lamin A mutant (Hennekes and Nigg 1994). However, in the case of $p21^{ras}$, another extensively studied protein with a CAAX box, farnesylation alone was found to be insufficient to confer strong membrane binding properties (Hancock et al. 1990, 1991b). Thus, it is attractive to speculate that binding of lamins to the nuclear membrane may also require additional factors. In particular, it should not be tacitly assumed that farnesyl groups will interact directly with membrane phospholipids, and it is entirely possible that they may also interact with integral membrane proteins (for discussion see Clarke 1992; Marshall 1993).

In this context, it is interesting that several integral membrane proteins have recently been shown to interact with nuclear lamins (e.g., Worman et al. 1988, 1990; Bailer et al. 1991; Simos and Georgatos 1992; Foisner and Gerace 1993). At least one of these proteins, the p54/p58 "lamin B receptor", contains several putative transmembrane domains (Worman et al. 1990) that might conceivably form proteinaceous hydrophobic pockets into which lamin-attached farnesyl-groups might intercalate (see Fig. 3). If this protein were indeed to interact with the hydrophobically modified C-terminus of lamins, its apparent specificity for B-type lamins would readily be explained. In apparent conflict with this model, it has been reported that lamin B and lamin B-receptors fractionate with different populations of mitotic membrane vesicles (Chaudhary and Courvalin 1993); however, other workers have recently obtained results indicating that B-type lamins and the p54/p58 protein remain associated throughout the cell cycle (Meier and Georgatos 1994). Perhaps it depends on the exact experimental conditions to what extent the interactions between these proteins are preserved in vitro. In particular, there is evidence that phosphorylation of the lamin B

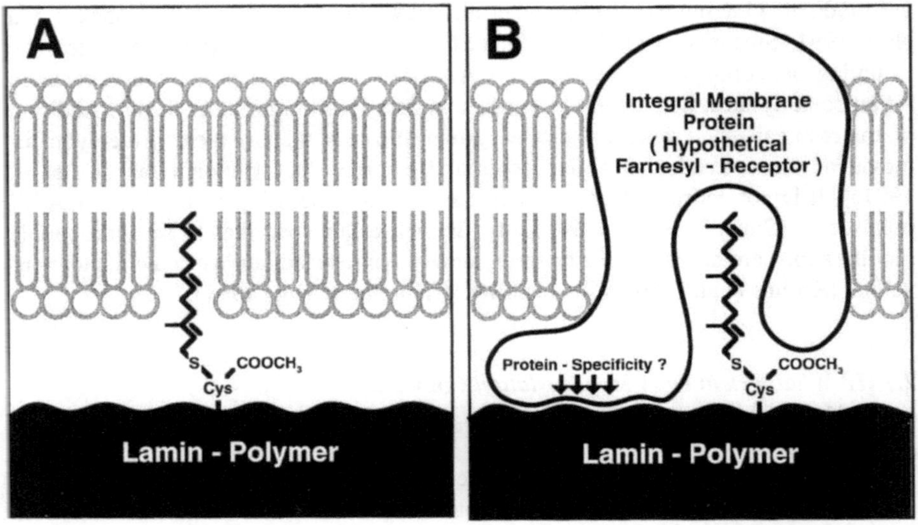

Fig. 3 A, B. Two alternative models describing the interaction of isoprenylated proteins with biological membranes. According to model **A**, the isoprenoid substitutent would intercalate directly into the phospholipid bilayer. (In view of the hydrophobic nature of the isoprene molecule, this model is plausible. Nevertheless, it should be noted that the methyl groups sticking out from the molecule's backbone might severely disturb the arrangement of the phospholipid fatty acids.) Alternatively, one could immagine that integral membrane proteins could accomodate isoprenoid substitutents within hydrophobic pockets (model **B**). Such pockets might easily be formed by protein oligomers and/or by proteins containing multiple membrane spanning domains (such as the p54/p58 lamin B receptor; Worman et al. 1990). At present, there are no definitive experimental data to prove or discard either model **A** or **B** (for further discussion see Clarke 1992; Epand et al. 1993; Marshall 1993). Note, however, that model **B** offers one important and attractive feature: protein-protein interactions, occuring in addition to protein-isoprene interactions, could explain why different isoprenylated proteins interact specifically with different types of membranes. (For earlier discussion of such models see also Nigg et al. 1992)

receptor may weaken its interactions with lamins during mitosis (Bailer et al. 1991; Courvalin et al. 1992). Thus, although lamins may remain bound to integral membrane proteins throughout the cell cycle in vivo, such complexes might easily be disrupted upon fractionation of mitotic cells in vitro. To resolve these issues, it may be necessary to functionally reconstitute the lamin B-receptor into artificial lipid bilayer systems.

To generate a tool for further analysis of the p54/p58 lamin B-receptor protein, a highly specific antibody was recently raised in our laboratory against the N-terminal 204 amino acids of the chicken lamin B receptor. This protein was expressed from a cDNA in *E. coli*, and purified to more than 95% purity by ammonium sulfate precipitation and cation exchange chromatography (H. Hennekes, S. Bailer, and E.A. Nigg, unpubl. results). The resulting rabbit antiserum (R24) recognizes p54 from chicken, mouse, and man (H. Hennekes and E.A. Nigg, unpubl. results), as well as from *Xenopus laevis* (Lourim and Krohne 1993; M. Lohka, pers. comm.). It can be used for immunoblotting, immunoprecipitation, and immunofluorescent staining experiments and should thus be applicable to a wide variety of studies on the biological function of the lamin B-receptor.

2.4 A Farnesylation-Independent Pathway for Targeting Lamins to the Nuclear Envelope

If, as outlined above, farnesylation and carboxyl-methylation of the CAAX box are required for targeting lamins to the nuclear membrane, then two obvious questions arise: first, how can lamin C reach the nuclear envelope despite the fact that it is synthesized without a CAAX box? Second, how can mature lamin A be re-utilized after mitosis, considering that it has lost the modified C-terminus due to proteolysis? The answer to these questions probably comprises two aspects. On the one hand, newly synthesized lamins A/C may conceivably be imported into the nucleus as heterodimers, in which case the C-terminus of the lamin A partner would carry the necessary hydrophobic modifications. On the other hand, there undoubtedly exists a farnesylation-independent pathway for lamina assembly (for review see Nigg et al. 1992): this pathway, however, is functional only during late stages of mitosis, when lamins have access to the surface of decondensing chromosomes before the nuclear envelope reforms. In direct support of this view, a recent study has revealed that integration of ectopically expressed lamin C into the nuclear lamina of embryonal carcinoma cells (which lack endogenous A/C lamins) occurs only after passage of such cells through mitosis (Horton et al. 1992) Similarly, lamina incorporation of CAAX box mutants of lamin A (i.e., mutants that could not be isoprenylated) required that cells pass through mitosis (Hennekes and Nigg 1994), and direct microinjection of unmodified lamin A into the nuclei of mammalian cells resulted in a very slow integration of such protein into the preexisting lamina (Goldman et al. 1992).

The precise order of events occurring during reassembly of the nuclear lamina and reformation of the nuclear envelope at the end of mitosis continues to be a subject of some debate (for discussion see Nigg et al. 1992; Pfaller and Newport 1992; Foisner and Gerace 1993). According to an early model, lamins of the A/C-type were proposed to bind first to the decondensing chromatin, before lamin B-containing vesicles associate and fuse at the chromosome surface (Burke and Gerace 1986). In support of this model, the chromatin-binding capacity of lamin A has been convincingly demonstrated, at least in vitro (Glass and Gerace 1990; Burke 1990; Höger et al. 1991; Yuan et al. 1991; Glass et al. 1993). Also, recent experiments have allowed the distinction of nuclear envelope-derived vesicles of different sizes with different protein contents, and it has been shown that these different vesicles reassemble at the chromosome surface at different times (Chaudhary and Courvalin 1993; Lourim and Krohne 1993; Vigers and Lohka 1992). Although the above model is attractive, it undoubtedly constitutes an oversimplification. In particular, it is clear that not all A/C lamins bind quantitatively to the chromatin surface, and that B-type lamins can also mediate envelope reformation in those cells that lack A/C lamins. Based on immunodepletion experiments carried out with *Xenopus* oocyte extracts, it has even been proposed that nuclear envelopes may reform around chromatin in the complete absence of lamins

(Newport et al. 1990). However, the interpretation of this particular study is complicated by the recent identification of an additional lamin isoform present in *Xenopus* oocytes (Lourim and Krohne 1993). Whatever the exact order of events taking place during reassembly of the nuclear envelope, and whatever the precise role of lamins, it seems clear that the formation of a nuclear lamina is required for the stability of the nuclear envelope. Also, re-formation of a lamina at the end of mitosis may be important for determining the spatial organization of interphase chromatin, and may thereby influence both the transcription and the replication of the genome during the subsequent cell cycle (for further discussion see Newport et al. 1990; Pfaller and Newport 1992; Nigg 1989, 1992a; Nigg et al. 1992; Foisner and Gerace 1993).

Acknowledgments. This work was supported by grants from the Swiss National Science Foundation (31-33615.92) and the Swiss Cancer League (FOR 205) to E.A.N., and a fellowship from the Deutsche Forschungsgemeinschaft to H.H.

References

Bailer SM, Eppenberger HM, Griffiths G & Nigg EA (1991) Characterization of a 54-kD protein of the inner nuclear membrane: evidence for cell cycle-dependent interaction with the nuclear lamina. J Cell Biol 114:389–400

Beven A, Guan Y, Peart J, Cooper C & Shaw P (1991) Monoclonal antibodies to plant nuclear matrix reveal intermediate filament-related components within the nucleus. J Cell Sci 98:293–302

Burke B & Gerace L (1986) A cell free system to study reassembly of the nuclear envelope at the end of mitosis. Cell 44:639–652

Burke B (1990) On the cell-free association of lamins A and C with metaphase chromosomes. Exp Cell Res 186:169–176

Chaudhary N & Courvalin JC (1993) Stepwise reassembly of the nuclear envelope at the end of mitosis. J Cell Biol 122:295–306

Clarke S (1992) Protein isoprenylation and methylation at carboxyl-terminal cysteine residues. Ann Rev Biochem 61:355–386

Courvalin JC, Segil N, Blobel G & Worman HJ (1992) The lamin B receptor of the inner nuclear membrane undergoes mitosis-specific phosphorylation and is a substrate for p34cdc2-type protein kinase. J Biol Chem 267:19035–19038

Dodemont H, Riemer D & Weber K (1990) Structure of an invertebrate gene encoding cytoplasmic intermediate filament (IF) proteins: implications for the origin and the diversification of IF proteins. EMBO J 9:4083–4094

Döring V & Stick R (1990) Gene structure of nuclear lamin LIII of *Xenopus laevis*; a model for the evolution of IF proteins from a lamin-like ancestor. EMBO J 9:4073–4081

Eggert M, Radomski N, Linder D, Tripier D, Traub P & Jost E (1993) Identification of novel phosphorylation sites in murine A-type lamins. Eur J Biochem 213:659–671

Epand RF, Xue CB, Wang SH, Naider F, Becker JM & Epand RM (1993) Role of prenylation in the interaction of the a-factor mating pheromone with phospholipid bilayers. Biochemistry 32:8368–8373

Foisner R & Gerace L (1993) Integral membrane proteins of the nuclear envelope interact with lamins and chromosomes, and binding is modulated by mitotic phosphorylation. Cell 73:1267–1279

Fuchs E & Weber K (1994) Intermediate filaments: structure, dynamics, function and disease. Ann Rev Biochem 63:345–382

Furukawa K & Hotta Y (1993) cDNA cloning of a germ cell specific lamin B3 from mouse spermatocytes and analysis of its function by ectopic expression in somatic cells. EMBO J 12:97–106

Gerace L & Blobel G (1980) The nuclear envelope lamina is reversibly depolymerized during mitosis. Cell 19:277–287

Gerace L & Burke B (1988) Functional organization of the nuclear envelope. Annu Rev Cell Biol 4:335–374

Glass JR & Gerace L (1990) Lamins A and B bind and assemble at the surface of mitotic chromosomes. J Cell Biol 111:1047–1057

Glass CA, Glass JR, Taniura H, Hasel KW, Blevitt JM & Gerace L (1993) The alpha-helical rod domain of human lamins A and C contains a chromatin binding site. EMBO J 12:4413–4424

Goldman AE, Moir RD, Montag Lowy M, Stewart M & Goldman RD (1992) Pathway of incorporation of microinjected lamin A into the nuclear envelope. J Cell Biol 119:725–735

Hancock JF, Paterson H & Marshall CJ (1990) A polybasic domain or palmitoylation is required in addition to the CAAX motif to localize p21ras to the plasma membrane. Cell 63:133–139

Hancock JF, Cadwallader K & Marshall CJ (1991a) Methylation and proteolysis are essential for efficient membrane binding of prenylated p21 K-ras(B). EMBO J 10:641–646

Hancock JF, Cadwallader K, Paterson H & Marshall CJ (1991b) A CAAX or a CAAL motif and a second signal are sufficient for plasma membrane targeting of ras proteins. EMBO J 10:4033–4039

Heald R & McKeon F (1990) Mutations of phosphorylation sites in lamin A that prevent nuclear lamina disassembly in mitosis. Cell 61:579–589

Heins S & Aebi U (1993) Making heads and tails of intermediate filament assembly dynamics and networks. Curr Opin Cell Biol 6:25–33

Hennekes H, Peter M, Weber K & Nigg EA (1993) Phosphorylation on protein kinase C sites inhibits nuclear import of lamin B2. J Cell Biol 120:1293–1304

Hennekes H & Nigg EA (1994) The role of isoprenylation in membrane attachment of nuclear lamins – a single point mutation prevents proteolytic cleavage of the lamin A precursor and confers membrane binding properties. J Cell Sci 107:1019–1029

Höger TH, Krohne G & Kleinschmidt JA (1991) Interaction of Xenopus lamins A and LII with chromatin in vitro mediated by a sequence element in the carboxyterminal domain. Exp Cell Res 197:280–289

Holtz D, Tanaka RA, Hartwig J & McKeon F (1989) The CaaX motif of lamin A functions in conjunction with the nuclear localization signal to target assembly to the nuclear envelope. Cell 59:969–977

Hornbeck P, Huang KP & Paul WE (1988) Lamin B is rapidly phosphorylated in lymphocytes after activation of protein kinase C. Proc Natl Acad Sci USA 85:2279–2283

Horton H, McMorrow I & Burke B (1992) Independent expression and assembly properties of heterologous lamins A and C in murine embryonal carcinomas. Eur J Cell Biol 57:172–183

Hovecar BA, Burns DJ & Fields AP (1993) Identification of protein kinase C (PKC) phosphorylation sites on human lamin B. Potential role of PKC in nuclear lamina structural dynamics. J Biol Chem 268:7545–7552

Kitten GT & Nigg EA (1991) The CaaX motif is required for isoprenylation, carboxyl methylation and nuclear membrane association of lamin B2. J Cell Biol 113:13–23

Krohne G, Waizenegger I & Höger TH (1989) The conserved carboxy-terminal cysteine of nuclear lamins is essential for lamin association with the nuclear envelope. J Cell Biol 109:2003–2011

Lehner CF, Kurer V, Eppenberger HM & Nigg EA (1986) The nuclear lamin protein family in higher vertebrates: identification of quantitatively minor lamin proteins by monoclonal antibodies. J Biol Chem 261:13293–13301

Lin F & Worman HJ (1993) Structural organization of the human gene encoding nuclear lamin A and nuclear lamin C. J Biol Chem 268:16321–16326

Loewinger L & McKeon F (1988) Mutations in the nuclear lamin proteins resulting in their aberrant assembly in the cytoplasm. EMBO J 7:2301–2309

Lourim D & Krohne G (1993) Membrane-associated lamins in *Xenopus* egg extracts: identification of two vesicle populations. J Cell Biol 123:501–512

Marshall CJ (1993) Protein prenylation: a mediator of protein-protein interactions. Science 259:1865–1866

Mattia E, Hoff WD, den Blaauwen J, Meijne AM, Stuurman N & van Renswoude J (1992) Induction of nuclear lamins A/C during in vitro-induced differentiation of F9 and P19 embryonal carcinoma cells. Exp Cell Res 203:449–455

McKeon F (1991) Nuclear lamin proteins: domains required for nuclear targeting, assembly, and cell-cycle-regulated dynamics. Curr Opin Cell Biol 3:82–86

McNulty AK & Saunders MJ (1992) Purification and immunological detection of pea nuclear intermediate filaments: evidence for plant nuclear lamins. J Cell Sci 103:407–414

Meier J & Georgatos SD (1994) Type B lamins remain associated with the integral envelope protein p58 during mitosis: implications for nuclear reassembly. EMBO J 13:1888–1898

Moir RD & Goldman RD (1993) Lamin dynamics. Curr Opin Cell Biol 5:408–411

Moss SB, Burnham BL & Bellve AR (1993) The differential expression of lamin epitopes during mouse spermatogenesis. Mol Reprod Dev 34:164–174

Newport JW, Wilson KL & Dunphy WG (1990) A lamin-independent pathway for nuclear envelope assembly. J Cell Biol 111:2247–2259

Nigg EA (1989) The nuclear envelope. Curr Opinions Cell Biol 1:435–440

Nigg EA (1991) The substrates of the cdc2 kinase. Seminars in Cell Biology 2:261–270

Nigg EA (1992a) Assembly and cell cycle dynamics of the nuclear lamina. Seminars in Cell Biology 3:245–253

Nigg EA (1992b) Assembly-disassembly of the nuclear lamina. Curr Opin Cell Biol 4:105–109

Nigg EA, Kitten GT & Vorburger K (1992) Targeting lamin proteins to the nuclear envelope: the role of *CAAX* box modifications. Biochemical Society Transactions 20:500–504

Ottaviano Y & Gerace L (1985) Phosphorylation of the nuclear lamins during interphase and mitosis. J Biol Chem 260:624–632

Peter M, Nakagawa J, Dorée M, Labbé JC & Nigg EA (1990) In vitro disassembly of the nuclear lamina and M-phase specific phosphorylation of lamins by cdc2 kinase. Cell 61:591–602

Peter M & Nigg EA (1991) Ectopic expression of an A-type lamin does not interfere with differentiation of lamin A-negative embryonal carcinoma cells. J Cell Sci 100:589–598

Peter M, Heitlinger E, Häner M, Aebi U & Nigg EA (1991) Disassembly of in vitro formed lamin head-to-tail polymers by cdc2 kinase. EMBO J 10:1535–1544

Peter M, Sanghera JS, Pelech SL & Nigg EA (1992) Mitogen-activated protein kinases phosphorylate nuclear lamins and display sequence specificity overlapping that of mitotic protein kinase p34cdc2. Eur J Biochem 205:287–294

Pfaller R & Newport JW (1992) Nuclear envelope assembly following mitosis. Meth Enzymol 219:60–72

Röber RA, Weber K & Osborn M (1989) Differential timing of nuclear lamin A/C expression in the various organs of the mouse embryo and the young animal: a developmental study. Development 105:365–378

Schafer WR & Rine J (1992) Protein prenylation – genes, enzymes, targets, and functions. Ann Rev Genet 26:209

Simos G & Georgatos SD (1992) The inner nuclear membrane protein p58 associates in vivo with a p58 kinase and the nuclear lamins. EMBO J 11:4027–4036

Sinensky M & Lutz RJ (1992) The prenylation of proteins. Bioessays 14:25–31

Sinensky M, Fantle K, Trujillo M, McLain T, Kupfer A & Dalton M (1994) The processing pathway of prelamin A. J Cell Sci 107:61–67

Smith A & Benavente R (1992) Identification of a short nuclear lamin protein selectively expressed during meiotic stages of rat spermatogenesis. Differentiation 52:55–60

Steinert PM, Marekow L, Fraser RDB & Parry DAD (1993) Keratin intermediate filament structure. J Mol Biol 230:436–452

Stewart M (1993) Intermediate filament structure and assembly. Curr Opin Cell Biol 5:3–11

Stick R, Angres B, Lehner CF & Nigg EA (1988) The fates of chicken nuclear lamin proteins during mitosis: evidence for a reversible redistribution of lamin B2 between inner nuclear membrane and elements of the endoplasmic reticulum. J Cell Biol 107:397–406

Vester B, Smith A, Krohne G & Benavente R (1993) Presence of a nuclear lamina in pachytene spermatocytes of the rat. J Cell Sci 104:557–563

Vigers GP, Lohka MJ (1992) Regulation of nuclear envelope precursor functions during cell division. J Cell Sci 102:273–284

Vorburger K, Lehner CF, Kitten G, Eppenberger HM & Nigg EA (1989) A second higher vertebrate B-type lamin: cDNA sequence determination and in vitro processing of chicken lamin B2. J Mol Biol 208:405–415

Ward GE & Kirschner MW (1990) Identification of cell cycle-regulated phosphorylation sites on nuclear lamin C. Cell 61:561–577

Weber K, Plessmann U & Ulrich W (1989) Cytoplasmic intermediate filament proteins of invertebrates are closer to nuclear lamins than are vertebrate intermediate filament proteins; sequence characterization of two muscle proteins of a nematode. EMBO J 8:3221–3227

Worman HJ, Yuan J, Blobel G & Georgatos SD (1988) A lamin B receptor in the nuclear envelope. Proc Natl Acad Sci USA 85:8531–8534

Worman HJ, Evans CD & Blobel G (1990) The lamin B receptor of the nuclear envelope inner membrane: a polytopic protein with eight potential transmembrane domains. J Cell Biol 111:1535–1542

Yuan J, Simons G, Blobel G & Georgatos SD (1991) Binding of lamin A to polynucleosomes. J Biol Chem 266:9211–9215

Towards a Molecular Understanding of Nuclear Pore Complex Structure and Function

N. PANTE [1], R. BASTOS[2], I. MCMORROW[2], K. N. GOLDIE[1], B. BURKE[2], and U. AEBI[2]

The nuclear pore complex (NPC) is a ~120 MDa (Reichelt et al. 1990) supramolecular assembly embedded in the double-membraned nuclear envelope (NE) that mediates nucleocytoplasmic transport in eukaryotic cells. Three-dimensional (3-D) reconstruction of both negatively stained (Hinshaw et al. 1992) and frozen-hydrated (Akey and Radermacher 1993) specimens have revealed the basic framework of the NPC. In addition, combination of different electron microscopies (EM), specimen preparation methods, and imaging techniques have yielded distinct peripheral NPC components such as the cytoplasmic and the nuclear ring, the cytoplasmic filaments and the nuclear basket (reviewed by Panté et al. 1993, 1994).

In contrast to the relatively large amount of structural studies, less is known about the chemical composition of the NPC. Based on its mass of about 120 MDa, it is estimated that the NPC may be built of multiple copies (i.e., 8 or 16, because of the 822 symmetry of the basic framework) of 100 or more different polypeptides. To date, roughly one dozen NPC proteins from diverse organisms have been cloned and sequenced, and 20–30 putative NPC proteins have been isolated or characterized. However, identification of most of these proteins with distinct NPC components remains elusive. Using colloidal gold-conjugated antibodies directed against a number of NPC proteins combined with quick freezing/freeze drying/rotary metal shadowing and embedding/thin sectioning, we have now localized two NPC polypeptides to distinct peripheral NPC components. A monoclonal antibody, termed QE5, recognized three NPC polypeptides, p250, NUP153, and p62 on Western blots (Fig. 1 a), and it labeled both the cytoplasmic and the nuclear face of *Xenopus* oocyte NEs. As documented in Fig. 1 b and d, on the cytoplasmic face of the NPC, gold-conjugated QE5 labeled (1) the cytoplasmic filaments, and (2) sites down in the pore disposed towards the nuclear opening. As can be depicted in Fig. 1c and d, on the nuclear face, the gold-conjugated QE5 labeled predominantly the nuclear baskets. To identify these distinct labeling sites with the different NPC proteins recognized by QE5, we used an anti-peptide antibody against NUP153 and a monospecific anti-p250 polyclonal antibody. As illustrated in Fig. 1 d–f, labeling with these two antibodies revealed that NUP153 is a constituent of the terminal ring of the nuclear basket, whereas p250 is a constituent of the cytoplasmic filament. In addition, QE5 immunoprecipitated three polypeptides, p75, p58, and p54, which were found to be associated with p250 (p75), and p62 (p58 and p54), respectively. The p62 complex was previously identified and characterized by Finlay et al. (1991).

[1] M. E. Müller Institute, Biozentrum, University of Basel, Switzerland.
[2] Department of Cell Biology, Harvard Medical School, Boston, USA.

45. Colloquium Mosbach 1994
The Cytoskeleton

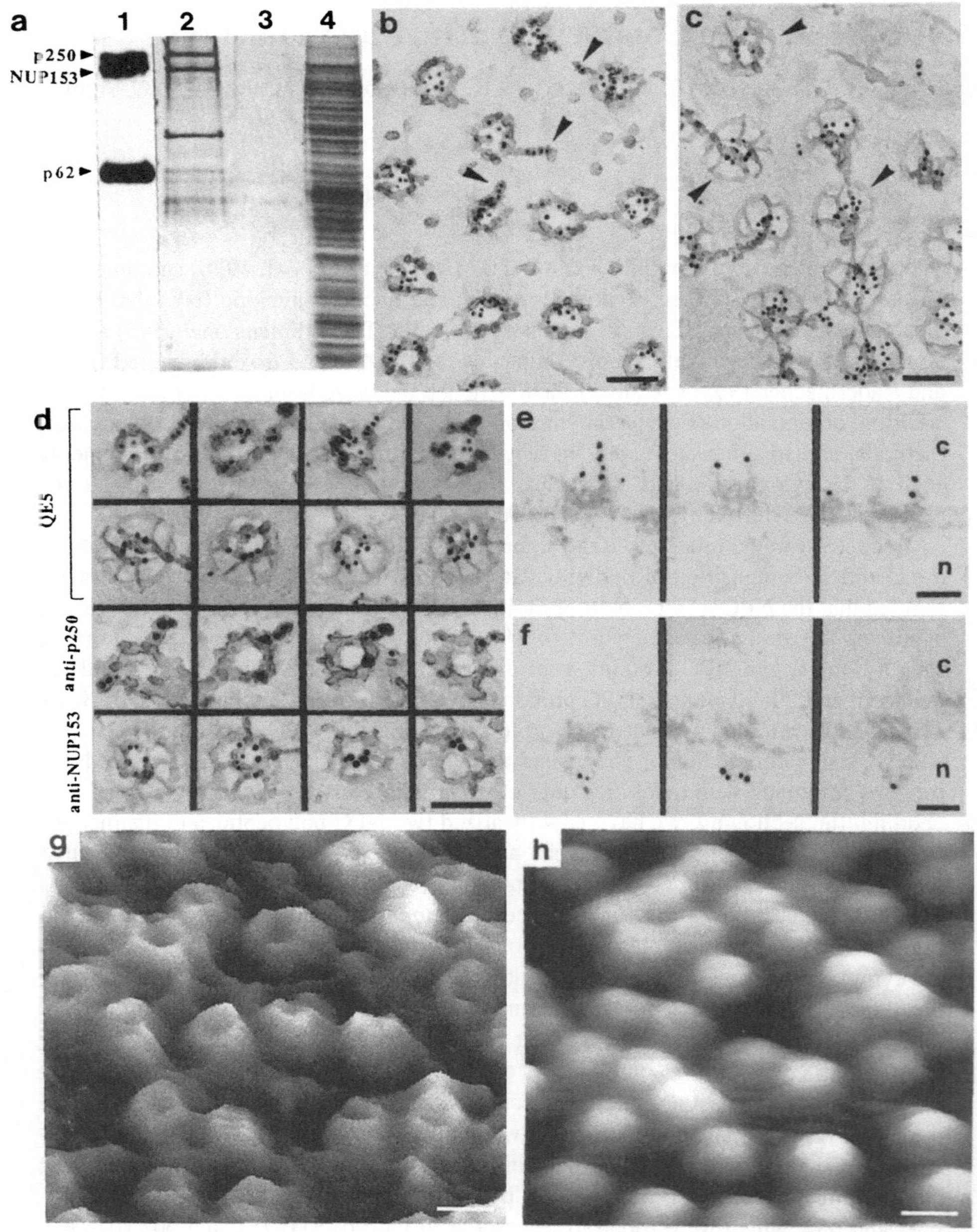
a
1
2
3
4
p250
NUP153
p62
b
c
d
QE5
anti-p250
anti-NUP153
e
c
n
f
c
n
g
h

Sometimes the "central pore" of the NPC is "plugged" with a massive, ~12 MDa particle called central plug, transporter, or central channel complex. Although this transporter has been recently reconstructed in 3-D (Akey and Radermacher 1993), it is still not clear to what extent it is an *integral* component of the NPC, or whether it represents at least in part material *in transit*. It is conceivable that a substantial fraction of what in projection appears as the central plug represents the terminal ring of the nuclear basket or material associated with it, which may have been squashed into the pore upon embedding the NPC in a thin layer of negative stain or a thin ice film. In support of this notion, the mass density of the central plug is attenuated significantly when isolated NEs or nuclei are depleted of divalent cations, a condition that causes opening of the nuclear baskets by dissociation of their terminal ring (Jarnik and Aebi 1991). Also, depending on the isolation and/or preparation conditions employed, both the abundance and appearance of the central plug are highly variable (Unwin and Milligan 1982; Akey 1990; Reichelt 1990; Jarnik and Aebi 1991). To investigate the nature and 3-D structure of the central channel complex more systematically, we have been employing a number of different buffer and incubation conditions and chemical fixation protocols. Accordingly, we have found that predominantly unplugged NPCs are yielded if during specimen preparation exogenous ATP is present, whereas the majority (~95%) of the NPCs harbor a massive central plug when the NEs are treated with Cu-orthophenanthroline, an oxidizing agent mediating S-S bridge formation.

Fig. 1. a Western blot analysis of rat liver nuclear envelopes (RLNE) with the monoclonal QE5 antibody. For this purpose, RLNEs were extracted with low-stringency TX buffer [50 mM triethanolamine (TEA), pH 7.4, 500 mM NaCl, 0,5% Triton X-100, 1 mM DTT, 1 mM PMSF], the extract (*lane 4*; revealed by silver staining) was immunoprecipitated with QE5, and the resulting immunoprecipitate was washed in low-stringency TX buffer. Samples were analyzed by SDS-PAGE and either immunoblotted with QE5 (*lane 1*) or silver stained (*lane 2*). For comparison, a silver stained, nonimmune control precipitate is shown in *lane 3*; 50x the amount used in *lane 4* was used for *lanes 1 to 3*. Accordingly, the monoclonal QE5 antibody immunoprecipitated p250, NUP153, p75, p62, p58, and p54 (*lane* 2), whereas only p250, NUP153, and p62 were detected by QE5 on Western blots (*lane 1*). Cytoplasmic (**b**) and nuclear (**c**) faces of quick-frozen/freeze-dried/rotary metal-shadowed spread *Xenopus* oocyte NEs labeled with the monoclonal QE5 antibody. QE5 that recognizes p250, NUP153 and p62 (**a**, *lane 1*), specifically labeled the cytoplasmic face of the NPC at (1) the cytoplasmic filaments (**b**, *arrowheads*), and (2) sites down in the pore disposed towards the nuclear opening. On the nuclear face, the gold-conjugated QE5 labeled predominantly the nuclear baskets (**c**, *arrowheads*). **d** Gallery of selected examples of quick-frozen/freeze-dried/rotary metal-shadowed NPCs labeled with the monoclonal QE5 antibody, a polyclonal antibody against p250, and an anti-peptide antibody against NUP153. The anti-p250 antibody exclusively labeled the cytoplasmic filaments, whereas the anti-NUP153 anti-peptide antibody exclusively labeled the terminal ring of the nuclear baskets. **e, f** Cross-sections and selected examples of Epon-embedded, Triton X-100-treated *Xenopus* oocyte nuclei labeled with anti-p250 antibody (**e**) and anti-NUP153 anti-peptide antibody (**f**). In agreement with the quick freezing/freeze drying/rotary metal shadowing results in **d**, the cross-sections revealed that the anti-p250 antibody exclusively labeled the cytoplasmic filaments, whereas the anti-NUP153 anti-peptide antibody exclusively labeled the terminal ring of the nuclear baskets. (*c* cytoplasmic, and *n* nuclear side of the nuclear envelope). Cytoplasmic (**g**) and nuclear (**h**) faces of *Xenopus* oocyte NEs kept in their native buffer environment while imaged with an atomic force microscope. Bar **b–h** 100 nm

Finally, to determine the native cytoplasmic and nuclear NPC topography, we have imaged NE-bound NPCs in their physiological buffer environment by atomic force microscopy (AFM). In agreement with EM of dehydrated specimens (Jarnik and Aebi 1991), corresponding AFM topographs revealed a high degree of asymmetry between the nuclear and cytoplasmic surface of the NPC. As documented in Fig. 1g and h, the cytoplasmic face of the NPC appears donut-like, whereas the nuclear face exhibits a dome-like appearance. In an effort to directly correlate NPC structure with function, we are trying to visualize different functional states of native NPCs by AFM after mechanical or chemical manipulation of spread NEs.

References

Akey CW (1990) Visualization of transport-related configurations of the nuclear pore transporter. Biophys J 58:341–355

Akey CW, Radermacher M (1993) Architecture of the *Xenopus* pore complex revealed by three-dimensional cryo-electron microscopy. J Cell Biol 122:1–19

Finlay DR, Meier E, Bradley P, Horecka J, Forbes DJ (1991) A complex of nuclear pore proteins required for pore function. J Cell Biol 114:169–183

Hinshaw JE, Carragher BO, Milligan RA (1992) Architecture and design of the nuclear pore complex. Cell 69:1133–1141

Jarnik M, Aebi U (1991) Towards a more complete 3-D structure of the nuclear pore complex. J Struct Biol 107:291–308

Panté N, Aebi U (1993) The nuclear pore complex. J Cell Biol 122:977–984

Panté N, Aebi U (1994) Towards understanding the 3-D structure of the nuclear pore complex at the molecular level. Curr Opin Struct Biol 4:187–196

Reichelt R, Holzenburg A, Buhle EL, Jarnik M, Engel A, Aebi U (1990) Correlation between structure and mass distribution of the nuclear pore complex, and of distinct pore complex components. J Cell Biol 110:883–894

Unwin PNT, Milligan RA (1982) A large particle associated with the perimeter of the nuclear pore complex. J Cell Biol 93:63–75

Titin and Nebulin: Giant Multitasking Protein Rulers in Muscle

K. WANG[1]

1 Muscle Activities and Cytoskeletons

When a striated muscle cell is stimulated, e.g. by nerve impulses, it is activated from the resting state. It develops contractile force, shortens, and then relengthens to its original dimension when stimulation ceases. The resting muscle, in the absence of any stimulation, is remarkably elastic when stretched and released. More than a century of muscle research has focused on the understanding of the structure and molecular processes which underlie active contraction. The widely accepted sliding filament model states that muscle develops active force by the cyclic attachment and detachment of myosin crossbridges to actin filaments and that muscle shortens when actin filaments are pulled to slide pass thick filaments, without changing the length of either filament (Huxley 1990). Despite major advances in the understanding of the molecular basis of active contraction, surprisingly little is known of how contracted muscle restores its length and how resting muscle responds to stretch and compression. It is also unclear how muscle cells manage to control the uniform and precise length of thick and thin filaments in the sarcomere. Recent studies of sarcomere-associated cytoskeletal lattices begin to shed light on both questions (reviewed by Wang 1985; Maruyama 1986, 1994; Price 1991; Trinick 1992).

The cytoplasm of striated muscle cells contains, besides actin and myosin filaments, a highly elaborate cytoskeletal network consisting of at least two biochemically distinct but interconnected lattices operating at two levels of structural hierarchy. At the cellular and organelle level, an intermediate filament lattice appears to envelop all sarcomeres linking the Z-lines and M-lines to the membrane skeleton (costamere), mitochondria, nuclei, and sarcoplasmic reticulum; at the sarcomere-myofilament level, a sarcomere matrix consisting of a set of elastic titin filaments and a set of inextensible nebulin filaments is thought to provide structural continuity and support of the sarcomere.

[1] Department of Chemistry and Biochemistry, Biochemical Institute and Cell Research Institute University of Texas at Austin, Austin, Texas 78712, USA. Phone and Fax: (512) 471-4065, E-mail: kuanw@uts.cc.utexas.edu.

45. Colloquium Mosbach 1994
The Cytoskeleton

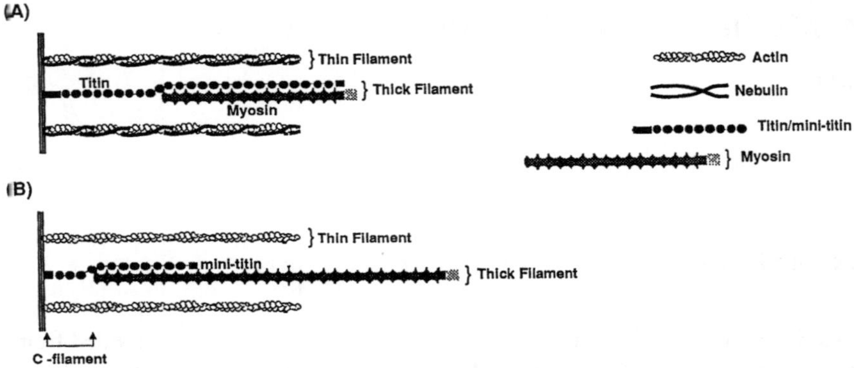

Fig. 1 A, B. Sarcomere models. **A** Vertebrate skeletal muscle containing both titin and nebulin. **B** Insect flight muscle containing minititin

2 Sarcomere Matrix: Titin and Nebulin as Molecular Filaments

Titin (also known as connectin) and nebulin, the major components of the sarcomere matrix, are both unprecedented giant proteins in the megadalton range and have been exceedingly intractable and challenging to characterize. It is now known that each protein represents a family of developmentally regulated modular proteins that vary in size in vertebrate striated muscles as well as invertebrate skeletal and smooth muscles. Their molecular properties are summarized in Tables 1 and 2 and new sarcomere models incorporating titin and nebulin are presented in Fig. 1.

3 Titin as a Protein Ruler of Thick Filaments

Titin, the largest protein known to date, is the third most abundant myofibrillar protein in vertebrate skeletal and cardiac muscles (Wang et al. 1979). In solution, native full-length titin (T_1, a-connectin) or its large fragments (T_2, β-connectin) is a long (up to 1 μm), slender, and flexible strand of beaded domains (3–4 nm in diameter), frequently with a larger globule at one end (Maruyama et al. 1984; Trinick et al. 1984; Wang et al. 1984; Nave et al. 1989). Its contour length coincides with the half sarcomere length of a resting muscle and is sufficient for a single titin molecule to extended half a sarcomere, traversing I-band and A-band with its N-terminus at the Z-line and its C-terminus at the M-line (Fürst et al. 1988; Labeit et al. 1992; Vinkemeier et al. 1993). The beaded morphology reflects the modular construction of titin sequence and domain motifs. The deduced protein sequence of titin indicates that titin is composed mainly of two types of sequence motifs (e.g. Labeit et al. 1992). One is related to fibronectin type-3 sequence (motif I) and the other to immunoglobulin C2 sequence (motif II). Both motifs are approximately 100 amino acids and fold indepen-

Table 1. Molecular properties of titin and nebulin

Properties	Titin filaments	Nebulin filaments
Skeletal muscle content	8–10%	2–3%
Cardiac muscle content	5–9%	0%
Subunit size	2.5–3.5 MDa	0.6–0.9 MDa
Size variant	Tissue- and development-specific	Tissue and development-specific
Morphology	Beaded string (4 nm diameter)	(Extended)
Sequence motif		
basic repeat	~100 Residues	~35 Residues
super-repeat	11 Modules near C-terminal half	7 Modules throughout
Protein domain	β-Barrel per repeat	α-Helix propensity
Contour length	0.9–1.2 μm, flexible	1.0–1.2 μm
Linear mass	2.7 ± 0.9 kDa/nm	?
Solubility	Favors high salt, high pH	?
Chromosome location	Human chromosome 2	Human chromosome 2
Actomyosin interaction	• Aggregates myosin rods • Aggregates actin filaments	• Cross-links actin-myosin • Inhibits actin sliding over myosin
	• Enhances actomyosin ATPase activity	• Inhibits actomyosin ATPase • Inhibition reversed by Ca/calmodulin
Protein interactions	Actin, AMP-deaminase, C-protein, M-protein, myomesin, myosin, X-protein	Actin, calmodulin, C-protein, myosin
Phosphorylation	• In vivo and in vitro • Multiple sites • Self-phosphorylating kinase	• In vivo and in vitro • Multiple sites
Self-assembly	• Oligomers • Helical aggregates	?
Homolog		
invertebrate muscles	Minititin, projectin, twitchin (~1 MDa)	?
nonmuscle	Titin-like proteins	

Table 2. Architecture and proposed functions of titin and nebulin filaments

Properties of function	Titin filaments	Nebulin filaments
Polypeptides per molecule	1 (or 2)	1 (or 2)
Molecules per filament	1	1
Filament ratios	6 (or 3) Filaments per half thick filament	4 (or 2) Filaments per thin filament
Axial lattice	• Each titin polypeptide spans from Z- to M-line (N- to C-terminus) • Titin filaments are intrinsically elastic along the entire length • In the sarcomere, titin filaments have two mechanical segments: a) Inextensible A-band segment b) Extensible I-band segment • The A-segment of titin interacts with external components of thick filaments	• Each nebulin polypeptide spans from Z-line to the end of thin filament (C- to N-terminus) • Inextensible
Radial lattice	Cofilament with myosin	Cofilament with actin
Resting tension	• Extension of I-segment in normal activity • Forced extension of A-segment beyond elastic limit	
Active tension	Modulate actomyosin interaction	Calmodulin-dependent regulation of actomyosin interaction
Sarcomere stability	Centering A-band	Stabilization of thin filaments
Length regulation	• Regulation of length and assembly of thick filaments • Matching super-repeat with thick filament periodicity	• Regulation of length and assembly of thin filaments • Matching super-repeat with thin filament periodicity
Myofibrillogenesis	• Template or scaffold for thick filaments • Docking A-band with IZI complex	• Template or scaffold for thin filaments • Assembly of IZI complex

hydrophobic core. Thus, a single, full-length titin molecule is made up of a string of ~300 β-barrels with their long axis roughly parallel to the string (Uchida et al. 1991). These domain motifs are building blocks of higher patterns along the molecule. Near the C-terminal half, a super-repeat consisting of a regular pattern of I-I-I-II-I-I-I-II-I-I-II occurs seven times. Such a super-repeat is thought to span the 43-nm periodicity of myosin heads along the thick filaments and serves as a length marker or a protein ruler that binds and regulates the length of thick filaments (Labeit et al. 1992). Indeed, titin and bacterially expressed motifs I and II bind to the myosin rod and several thick filament-associated proteins with high affinity (Maruyama et al. 1989; Nave et al. 1989; Koretz et al. 1993; Soteriou et al. 1993b). Interestingly, a myosin light chain kinase-like domain is found near the C-terminus and is localized to ~100 nm from the center of the M-line in situ (Labeit et al. 1982). This kinase-like domain is likely to be active, since titin self-phosphorylates and is phosphorylated in vivo (Somerville and Wang 1988; Takano-Ohmuro et al. 1992). The physiological significance of titin kinase activity and its native substrates remain a mystery.

4 Titin as Molecular Spring

The elasticity of titin filaments in situ has been predicted since the time of its discovery (Maruyama et al. 1977; Wang et al. 1979). Experimentally, this is demonstrated with immunoelectron microscopy by the manner with which unique titin epitopes translocate when the sarcomere is stretched. The epitopes within the I-band move away from both the Z-line and the M-line as if they are suspended on a rubber band, whereas epitopes in the A-band remain at a fixed distance from the M-line (Wang and Ramirez-Mitchell 1983; Itoh et al. 1988, Wang et al. 1993). However, when myosin filaments are disassembled by salt extraction, A-band epitopes then translocate elastically (Higuchi et al. 1992; Wang et al. 1993). These data indicate that the titin filament is intrinsically elastic along most of its length. However, due to its strong lateral association with the myosin filaments, the segment of titin that spans the A-band is prevented from stretching. The remaining segment outside of a the A-band is extensible and acts as an elastic connector linking thick filaments to the Z-line, thus imparting structural continuity of the sarcomere.

5 Elasticity of Muscle Cytoskeletons and Segmental Extension of Titin

How do the two interconnected muscle cytoskeletons behave structurally and mechanically when muscle cells change length and shape? If both are force-bearing lattices, what is the relative contribution of each lattice to the long-range elasticity? These questions are being explored by mechanical studies and immunoelectron microscopy of skinned or extracted muscle fibers.

The mechanical roles of titin have attracted much attention. Its contribution to sarcomere stability and long-range elasticity of the sarcomere was suggested by the ob-

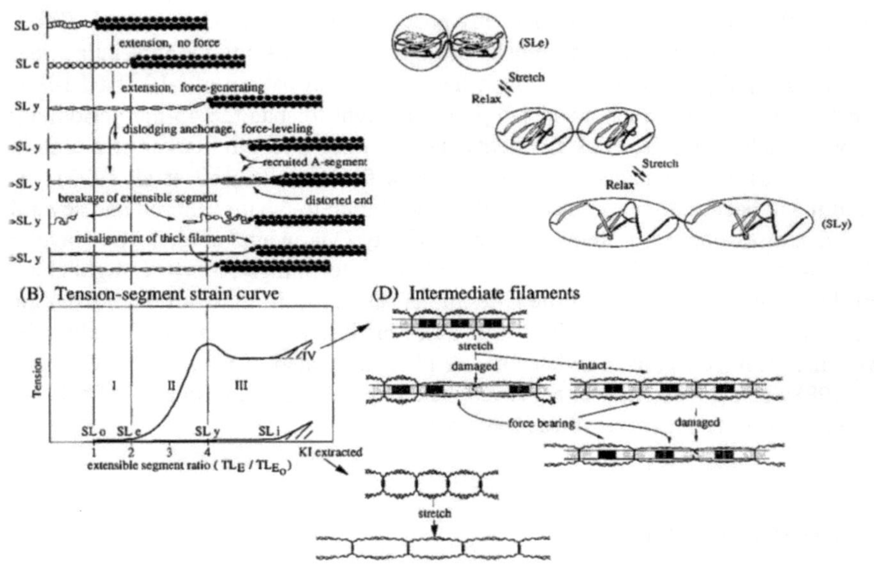

Fig. 2. Elasticity of muscle cytoskeletons. **A** Proposes structural changes in titin/myosin composite filaments upon stretch. **B** Resting tension-sarcomere length relationship of myofibrillar bundles containing both sarcomere matrix and intermediate filaments. Stretching of the extensible segment of titin/myosin composite filaments is likely to contribute exclusively to the exponential rise in tension (*zone II*, between *SLe* and *SLy*). The detachment of part of the anchored titin segment at and beyond the elastic limit of the sarcomere *(SLy)* contributes to the leveling of tension at higher degree of stretch *(zone III)*. Stretching of intermediate filament lattice contributes to the further increase in tension beyond 5 μm *(zone IV)*. The resting tension curves of several types of muscles with different titin isoforms and elastic limits can be normalized to the same master curve by plotting resting tension as function of the degree of stretch of the extensible segment of titin (extensible segmental ratio in **B**, see Wang et al. 1993 for details)

servations that when titin was destroyed preferentially by radiation or by proteolysis (Horowits et al. 1986; Higuchi 1992), resting tension diminished and thick filaments misaligned. Titin is thought to keep thick filaments centered in the sarcomere, ensuring the development of equal active force on both sides of the A-band (Horowits and Podolsky 1987). Its involvement in generating and transmitting resting tension was established by comparing the resting tension – sarcomere length relationships with the stretch behavior of titin revealed by epitope translocation studies. A positive correlation between sarcomere elasticity and titin extensibility in several types of skeletal muscles that express titin-size isoforms led us to propose a *segmental extension model of resting tension* (Wang et al. 1991; Granzier and Wang 1993a, b). As illustrated in Fig. 2, the key molecular event that generates resting tension appears to be a cyclic change in contour length of the titin segment in the I-band, perhaps by a reversible unfolding and folding of titin domains (Wang et al. 1991, 1993; Politou et al. 1994; Soteriou et al. 1993a). The tension rises exponentially upon stretching until it reaches a point where a portion of the anchored titin segment in the A-band is drawn out by the resting tension. Beyond this elastic limit (yield point) of the sarcomere, corresponding to a 300–400% extension of the titin segment, the tension levels off as the recruitment of titin molecules from the A-segment continues. The release of previously anchored titin appears to be an effective stress-management mechanism, since it prevents further rise in tension and limits the potential damages if breakage of titin and fracturing of the sarcomere occur in overly stretched muscles (Wang et al. 1993).

Mechanically, the titin/myosin composite filament can be viewed as a *dual-stage molecular spring* with the short extensible segment responding to normal stretches and the longer, anchored segment as a backup to handle extreme stretch (Wang et al. 1991, 1993; Granzier and Wang 1993a, b).

By studying muscles treated with KI to remove actomyosin and to dislodge titin from the M-line, the intermediate filament network preparation is shown to contribute to resting tension, but only beyond 5 μm sarcomere length (Wang et al. 1993). Thus, both sarcomere matrix and intermediate filaments are force-bearing elements of the cytoskeletal network.

6 Minititin and Regulation of Active Contraction

A group of smaller homologs of titin is found in insect flight muscles and in skeletal and smooth muscles of invertebrates. Known as minititin, twitchin, or projectin (Saide 1981; Benian et al. 1989; Nave et al. 1990, 1991), these ~1 Md proteins share with titin the beaded-string morphology, sequence motifs, and the presence of myosin light chain kinase domain near the C-terminus (Maroto et al. 1992). In the insect flight muscles, minititin (0.26 μm long) extends from the Z-line across the short I-band (this portion is termed C-filament; Fig. 1) and anchors near the outer region of the A-band, without reaching the M-line (Nave et al. 1990). This segmental structure of minititin gives rise to multiphasic resting tension-sarcomere length curves that can be explained with the segmental extension model originally developed for vertebrate skeletal muscles (Granzier and Wang 1993a).

Studies of insect flight muscles also revealed another functional manifestation of elastic filaments, i.e., their ability to regulate actomyosin interaction. For these muscles, actin-myosin interactions occur in both a resting state (as "weak" bridges that do not generate force) and in an activated state (strong bridges that do generate force). The striking finding is that the level of resting tension increases in direct proportion to the number of weak and strong bridges (Granzier and Wang 1993a). It is conceivable that stretched minititin causes strain or altered packing of myosin in the thick filaments such that both types of actomyosin interactions are favored. In other words, resting tension appears to be a prerequisite for the formation of cross-bridges (Fig. 3). Whether such interplays exist between elastic filaments and actin-myosin interaction in other types of muscles, such as cardiac muscles, is yet unclear.

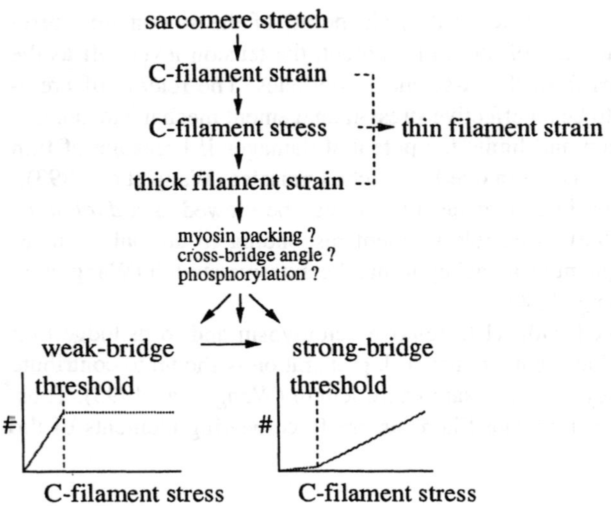

Fig. 3. Influence of resting tension on active contraction of insect flight muscle. Resting tension generated by the stretching of the extensible segment of minititin (C-filament) is thought to be transmitted to the thick filament. Structural changes in the thick filaments then promote the formation of both weak bridges (in resting state) and force-producing strong bridges (when subsequently activated); see Granzier and Wang (1993) for details

7 Nebulin as a Protein Ruler of Thin Filaments

Nebulin comprises 2–3% of the myofibrillar protein of skeletal muscle (Locker and Wild 1986; Wang and Wright 1988; Kruger et al. 1991; Labeit et al. 1991) and displays tissue- and development-specific isoforms, with a size ranging from 600 to 900 kDa in a variety of muscle types. Immunoelectron microscopy suggests that a single nebulin molecule spans the whole length of the thin filaments with its C-terminus anchored at the Z-line and that it moves along with actin filaments during muscle contracting (Wang and Wright 1988; Maruyama et al. 1989; Pierobon-Bormioli et al. 1989; Jin and Wang 1991; Kruger et al. 1991; Wright et al. 1993). These data suggest that nebulin forms a composite filament with actin/tropomyosin/troponin, perhaps by binding laterally to actin in situ. Significantly, the size of nebulin isoforms is proportional to the length of the thin filaments in many skeletal muscles (Kruger et al. 1991; Labeit et al. 1991). In contrast, cardiac muscles that lack nebulin display a high degree of variability in the lengths of the thin filaments (Robinson and Winegrad 1979). These correlations strongly suggest that nebulin may act as a protein ruler that regulates the length of thin filaments.

Although native nebulin has yet to be isolated and characterized, close to 20 kb of a human nebulin transcript has been isolated and sequenced (Wang et al. 1990). The deduced amino acid sequence of human nebulin shows a ~35 residue motif that is repeated approximately 200 times (Stedman et al. 1988; Zeviani et al. 1988; Wang et al. 1990; Labeit et al. 1991). This sequence also contains close to 25 copies of a super-repeat which consists of 7 of the 35 residue modules (Wang et al. 1990). It has been proposed that the 35-residue repeat is the basic structural unit of the actin-binding domains in nebulin (Jin and Wang 1991; Kruger et al. 1991; Labeit et al. 1991). Actin-binding studies with recombinant nebulin fragments containing 2 to 15 modules, small native nebulin fragments, and synthetic peptides confirmed this prediction and indicate that nebulin may contain a string of 200 actin-binding domains along its length (Jin and Wang 1991; Tatsumi et al. 1992; Chen et al. 1993; Pfuhl et al. 1994). If all sites are operative in situ, the nebulin would act as a zipper in its lateral association with actin (Chen et al. 1993). A simple matching between one nebulin module with one actin subunit and one super-repeat with the 38-nm filament periodicity would allow nebulin to operate as a protein ruler to determine or stabilize the length of actin filaments (Kruger et al. 1991; Labeit et al. 1991).

It is interesting that the C-terminal 200 residues of nebulin are completely devoid of the 35-residue repeating module, but does contain a src homology 3 (SH3) domain (Wang et al. 1990). Since the C-terminus of nebulin is anchored at the Z-line (Wright et al. 1993), the SH3 domain way represent a site of attachment to the Z-line. It is thought that SH3 motifs in cytoskeletal proteins enhance their potential to associate with cell membranes (Musacchio et al. 1992). The attachment of nebulin SH3 to the Z-line suggests that SH3 motifs may blind to actin or proteins that are enriched in actin filament attachment sites. Z-line proteins, such as α-actinin (Nave et al. 1990), titin, and Cap-Z (Caldwell et al. 1989), are potential anchor proteins.

8 Nebulin as a Calcium/Calmodulin-Mediated Regulatory Protein of Active Contraction

Recent studies on the effect of nebulin fragments on actin-myosin interaction and its regulation by calmodulin raise the intriguing possibility that nebulin might have regulatory functions on active contraction (Root and Wang 1994). Nebulin fragments bind with high affinity to actin, myosin, and myosin head. Nebulin fragments from the N-terminal half that are situated in the actomyosin overlap region of the sarcomere inhibit actomyosin ATPase activities as well as sliding velocities of actin over myosin in in vitro motility assays; while a nebulin fragment near the C-terminus which is localized to the Z-line does not prevent actin sliding. Significantly, calmodulin reverses the inhibition of ATPase and accelerates actin sliding in a calcium-dependent manner. Calmodulin with calcium greatly reduces the binding of nebulin fragments to both actin and myosin. The data suggest that the nebulin-calmodulin system is reminiscent of caldesmon and calponin in smooth muscles (reviewed in Matsumura and Yamashiro 1993) and may function as an calcium-linked regulatory system that is distinct from tropomyosin/troponin on the actin filaments of skeletal muscles. Nebulin may hold myosin heads close to actin in an orientation that prevents random interaction in resting muscles yet facilitates cross-bridge cycling upon activation by calcium and calmodulin.

9 Titin and Nebulin in Developing Muscles

Titin also plays significant roles in the assembly of myofibrils in developing muscles (reviewed in Fulton and Isaacs 1991; Epstein and Fischman 1991; Stromer 1992). It becomes detectable early in postmitotic mononucleated myoblasts (Hill et al. 1986; Colley et al. 1990) and accumulates and assembles in multinucleated myotubes. The reported sequence of events varies somewhat in different embryonic and cultured cells of skeletal and cardiac muscles (Tokuyasu and Maher 1987; Wang et al. 1988; Fürst et al. 1989; Terai et al. 1989; Schultheiss et al. 1990; Handel et al. 1989, 1991). In some studies, titin distributes first diffusely in the cytoplasm and then one end of it assembles with actin and α-actinin into primitive IZI complexes that appear as random patches. Subsequently, the other end of titin links up with thick filament bundles to form periodic, sarcomere-like arrays (Schultheiss et al. 1990). These data implicate titin in the assembly of IZI complexes and in the docking IZI complexes with the A-band. In others, titin and myosin always codistribute and can be cross-linked together at the onset of their expression (Hill et al. 1986; Fürst et al. 1989; Terai et al. 1989; Isaacs et al. 1992), consistent with a strong interaction between titin and myosin. Such studies emphasize instead its potential role as a template for thick filament assembly. A recent report on the association of titin with intermediate filaments along stress fibers in human skeletal muscle cells (van der Ven et al. 1993) supports its additional involvement in the architecture of other cytoskeletal filaments as well.

The incorporation of nebulin follows a pattern that is similar to many thin filament proteins: a diffuse cytoplasmic distribution, followed by its incorporation into the nascent stress-fiber-like structures, and finally the mature striated distribution (Itoh et al. 1988; Fürst et al. 1989; Komiyama et al. 1992; van der Ven et al. 1993). During maturation, nebulin staining changes from single stripes containing the nascent Z-line to doublets around the Z-line during maturation. It is unclear whether this change reflects the expression of new isoforms or a reorganization of the preexisting isoforms. The exact timing of nebulin incorporation may vary among species (van der Ven et al. 1993). In chicken skeletal muscle cells, nebulin assembles into the IZI complex after α-actinin incorporation, but prior to the assembly of myosin and titin into the nascent striated myofibril (Komiyama et al. 1992). Whether nebulin acts as a template to direct the assembly of actin filaments in the IZI complex or as a reinforcing scaffold to stabilize existing actin filaments remains to be determined.

10 Giant Proteins and Ultrathin Filaments in Smooth Muscle and Nonmuscle Cells

Titin-like proteins and ultrathin filaments (2–5 nm) are increasingly reported in nonmuscle and smooth muscle cells (sea urchin eggs, Pudles et al. 1990; chicken intestinal epithelial cells, Eilertsen and Keller 1992; molluscan smooth muscle cells, Vibert et al. 1993). What has emerged is the realization that titin may represent a newly identified cytoskeleton that is thin, elastic, and widespread in nature (reviewed in

Roberts 1987). In contrast to the well-characterized major cytoskeletal filaments (microfilaments, microtubules, and intermediate filaments) that are assembled noncovalently on demand from small building blocks, a fundamentally distinct feature of this elastic cytoskeleton is that these giant, elongated molecules are preassembled covalently during synthesis, with uniform length and a fixed order of functional domains. Unique structural and functional features are thus to be expected. The potential contribution of such a cytoskeleton in the cytoplasmic organization and function of smooth muscle and nonmuscle cells remains a fertile ground for future exploration.

Acknowledgments. I deeply appreciate my present and past colleagues and collaborators for their contributions to the works that are described in this chapter. This work is dedicated to the memory of my esteemed colleague, Ruben Ramirez-Mitchell, who fought a courageous battle with cancer. His dedication, friendship, and spirit will be sorely missed.

References

Benian GM, Kiff JE, Neckelmann N, Moerman DG, Waterston RH (1989) Sequence of an unusually large protein implicated in regulation of myosin activity in *C. elegans*. Nature 342:45–50

Caldwell JE, Heiss SG, Mermall V, Cooper JA (1989) Effects of CapZ, an actin capping protein of muscle, on the polymerization of actin. Biochemistry 28:8506–8514

Chen M-JG, Shih C-L, Wang K (1993) Nebulin as an actin zipper. J Biol Chem 268:20327–20334

Colley NJ, Tokuyasu KT, Singer SJ (1990) The early expression of myofibrillar proteins in rounded postmitotic myoblasts of embryonic skeletal muscle. J Cell Sci 95:11–22

Eilertsen KJ, Keller TCS (1992) Identification and characterization of two huge protein components of the brush border cytoskeleton: evidence for a cellular isoform of titin. J Cell Biol 119:549–557

Epstein HF, Fischman DA (1991) The molecular basis of protein assembly in muscle development. Science 251:1039–1044

Fulton AB, Isaacs WB (1991) Titin, a huge, elastic sarcomeric protein with a probable role in morphogenesis. BioEssays 13:157–161

Fürst DO, Osborn M, Nave R, Weber K (1988) The organization of titin filaments in the half sarcomere revealed by monoclonal antibodies in immunoelectron microscopy: a map of ten nonrepetitive epitopes starting at the Z line extends close to the M line. J Cell Biol 106:1563–1572

Fürst DO, Osborn M, Weber K (1989) Myogenesis in the mouse embryo: differential onset of expression of myogenic proteins and the involvement of titin in myofibril assembly. J Cell Biol 109:517–527

Granzier HLM, Wang K (1993a) Interplay between passive tension and strong and weak binding cross-bridges in insect indirect flight muscle. J Gen Physiol 101:235–270

Granzier HLM, Wang K (1993b) Passive tension and stiffness of vertebrate skeletal and insect flight muscles: the contribution of weak cross-bridges and elastic filaments. Biophys J 65:2141–2159

Handel SE, Greaser ML, Schultz E, Wang S-M, Bulinski JC, Lin JJ, Lessard J (1991) Chicken cardiac myofibrillogenesis: studies with antibodies specific for titin and the muscle and nonmuscle isoforms of actin and tropomyosin. Cell Tissue Res 263:419–430

Handel SE, Wang SM, Greaser ML, Schultz E, Bulinski JC, Lessard JL (1989) Skeletal muscle myofibrillogenesis as revealed with a monoclonal antibody to titin in combination with detection of the α- and γ-isoforms of actin. Dev Biol 132:35–44

Higuchi H (1992) Changes in contractile properties with selective digestion of connectin (titin) in skinned fibers of frog skeletal muscle. J Biochem 111:291–295

Higuchi H, Suzuki K, Yoshioka T, Maruyama K, Umazume Y (1992) Localization and elasticity of connectin (titin) filaments in skinned frog muscle fibres subjected to partial depolymerization of thick filaments. J Mus Res Cell Motil 13:285–294

Hill CS, Duran S, Lin Z, Weber K, Holtzer H (1986) Titin and myosin, but not desmin, are linked during myofibrillogenesis in postmitotic mononucleated myoblasts. J Cell Biol 103:2185–2196

Horowits R, Podolsky RJ (1987) The positional stability of thick filaments in activated skeletal muscle depends on sarcomere length: evidence for the role of titin filaments. J Cell Biol 105:2217–2223

Horowits R, Kempner ES, Bisher ME, Podolsky RJ (1986) A physiological role for titin and nebulin in skeletal muscle. Nature 323:160–164

Huxley HE (1990) Sliding filaments and molecular motile systems. J Biol Chem 265:8347–8350

Isaacs WB, Kim IS, Struve A, Fulton A (1992) Biosynthesis of titin and myosin heavy chain in developing skeletal muscle. Pro Natl Acad Sci USA 89:7496–7500

Itoh Y, Suzuki T, Kimura S, Ohashi K, Higuchi H, Sawada H, Shimizu T, Shibata M, Maruyama K (1988) Extensible and less-extensible domains of connectin filaments in stretched vertebrate skeletal muscle sarcomeres as detected by immunofluorescence and immunoelectron microscopy using monoclonal antibodies. J Biochem 104:504–508

Jin J-P, Wang K (1991) Cloning, expression and protein interaction of human nebulin fragments composed of varying numbers of sequence modules. J Biol Chem 266:21215–21223

Komiyama M, Zhou Z-H, Maruyama K, Shimada Y (1992) Spatial relationship of nebulin relative to other myofibrillar proteins during myogenesis in embryonic chick skeletal muscle cells in vitro. J Mus Res Cell Motil 13:48–54

Koretz JF, Irving TC, Wang K (1993) Filamentous aggregates of native titin and binding of C-protein and AMP-deaminase. Arch Biochem Biophys 304:305–309

Kruger M, Wright J, Wang K (1991) Nebulin as a length regulator of thin filaments of vertebrate skeletal muscles: correlation of thin filament length, nebulin size and epitope profile. J Cell Biol 115:97–107

Labeit S, Gautel M, Lakey A, Trinick J (1992) Towards a molecular understanding of titin. EMBO J 11:1711–1716

Labeit S, Gibson T, Lakey A, Leonard K, Zeviani M, Knight P, Wardale J, Trinick J (1991) Evidence that nebulin is a protein-ruler in muscle thin filaments. FEBS Lett 282:313–316

Locker RH, Wild DJC (1986) A comparative study of high molecular weight proteins in various types of muscles across the animal kingdom. J Biochem 99:1473–1484

Maroto M, Vinos J, Marco R, Cervera M (1992) Autophosphorylating protein kinase activity in titin-like arthropod projectin. J Mol Biol 224:287–291

Maruyama K (1986) Connectin, an elastic filamentous protein of striated muscle. In Rev Cytol 104:81–114

Maruyama K (1994) Connectin, an elastic protein of striated muscle. Biophys Chem 50:73–85

Maruyama K, Matsubara S, Natori R, Nonomura Y, Kimura S, Ohashi K, Murakami F, Handa F, Eguchi G (1977) Connectin, an elastic protein of muscle. J Biochem 82:317–337

Maruyama K, Kimura S, Yoshidomi H, Sawada H, Kikuchi M (1984) Molecular size and shape of β-connectin, an elastic protein of striated muscle. J Biochem 95:1423–1433

Maruyama T, Nakauchi Y, Kimura S, Maruyama K (1989) Binding of connectin to myosin filaments. J Biochem 105:323–326

Matsumura F, Yamashiro S (1993) Caldesmon. Curr Opinion Cell Biol 5:70–76

Musacchio A, Gibson T, Lehto V-P, Saraste M (1992) SH3 – an abundant protein domain in search of a function. FEBS Lett 307:55–61

Nave R, Fürst DO, Weber K (1989) Visualization of the polarity of isolated titin molecules: a singular globular head on a long rod as the M band anchoring domain? J Cell Biol 109:2177–2187

Nave R, Fürst DO, Weber K (1990) Interaction of α-actinin and nebulin in vitro: support for the existence of a fourth filament system in skeletal muscle. FEBS Lett 269:163–166

Nave R, Fürst D, Vinkemeier U, Weber K (1991) Purification and physical properties of nematode mini-titins and their relation to twitchin. J Cell Sci 98:491–496

Pfuhl M, Winder SJ, Pastore A (1994) Nebulin, a helical actin binding protein. EMBO J 13:1782–1789

Pierobon-Bormioli S, Betto R, Salviati G (1989) The organization of titin (connectin) and nebulin in the sarcomeres: an immunocytolocalization study. J Mus Cell Mot 10:446–456

Politou AS, Gauel M, Pfuhl M, Labeit S, Pastore A (1994) Immunoglobulin-type domains of titin: same fold, different stability. Biochemistry 33:4730–4737

Price MG (1991) Striated muscle endosarcomeric and exosarcomeric lattices. Adv Struct Biol 1:175–208

Pudles J, Mondjou M, Hisanaga S-I, Maruyama K, Sakai H (1990) Isolation, characterization and immunochemical properties of a giant protein from sea urchin egg cytomatrix. Exp Cell Res 189:253–260

Roberts TM (1987) Fine (2-5-nm) filaments: new types of cytoskeletal structures.

Robinson RF, Winegrad S (1979) The measurement and dynamic implications of thin filament lengths in heart muscle. J Physiol 286:607–619

Root D, Wang K (1994) Calmodulin sensitive interaction of human nebulin fragments with actin and myosin. Biochemistry 33:12581–12591

Saide JD (1981) Identification of a connecting filament protein in insect fibrillar flight muscle. J Mol Biol 153:661–679

Schultheiss T, Lin Z, Lu MH, Murray J, Fischman DA, Weber K, Masaki T, Imamura M, Holtzer H (1990) Differential distribution of subsets of myofibrillar protein in cardiac nonstriated and striated myofibrils. J Cell Biol 110:1159–1172

Somerville L, Wang K (1988) Sarcomere matrix of striated muscle: in vivo phosphorylation of titin and nebulin in mouse diaphragm muscle. Arch Biochem Biophys 262:118–129

Soteriou A, Clarke A, Martin S, Trinick J (1993a) Titin folding energy and elasticity. Proc R Soc Lond B 254:83–86

Soteriou A, Gamage M, Trinick J (1993b) A survey of interactions made by the giant protein titin. J Cell Sci 104:119–123

Stedman H, Browning K, Oliver N, Oronzi-Scott M, Fischbeck K, Sarkar S, Sylvester J, Schmickel R, Wang K (1988) Nebulin cDNAs detect a 25-kilobase transcript in skeletal muscle and localize to human chromosome 2. Genomics 2:1–7

Stromer MH (1992) Immunocytochemical localization of proteins in striated muscle. Int Rev Cytol 142:61–144

Takano-Ohmuro H, Nakauchi Y, Kimura S, Maruyama K (1992) Autophosphorylation of β-connectin (titin 2) in vitro. Biochem Biophys Res Commun 183:31–35

Tatsumi R, Hattori A, Takahashi K (1992) Purification and characterization of nebulin subfragments produced by 0.1 mM $CaCl_2$. J Biochem 112:780–785

Terai M, Komiyana M, Shimada Y (1989) Myofibril assembly is linked with vinculin, actinin and cell substrate contacts in embryonic cardiac myocytes in vitro. Cell Mot Cytoskel 12:185–191

Tokuyasu KT, Maher PA (1987) Immunocytochemical studies of cardiac myofibrillogenesis in early chick embryos. II. Generation of α-actinin dots within titin spots at the time of the first myofibril formation. J Cell Biol 105:2795–2801

Trinick J (1992) Understanding the functions of titin and nebulin. FEBS Lett 307:44–48

Trinick J, Knight P, Whiting A (1984) Purification and properties of native titin. J Mol Biol 180:331–356

Uchida K, Harada I, Nakauchi Y, Maruyama K (1991) Structural properties of connectin studied by ultraviolet resonance Raman spectroscopy and infrared dichroism. FEBS Lett 296:35–38

van der Ven PEM, Schaart G, Croes HJE, Jap PHK, Ginsel LA (1993) Titin aggregates associated with intermediate filaments align along stress fiber-like structures during human skeletal muscle cell differentiation. J Cell Sci 106:749–759

Vibert P, Edelstein SM, Castellani L, Elliott BWJ (1993) Mini-titins in striated and smooth molluscan muscles: structure, location and immunological crossreactivity. J Mus Res Cell Motil 14:598–607

Vinkemeier U, Obermann W, Weber K, Fürst DO (1993) The globular head domain of titin extends into the center of the sarcomeric M band. J Cell Sci 106:319–330

Wang K (1985) Sarcomere-associated cytoskeletal lattices in striated muscle, pp. 315–369. In.: Shay JW (ed) Cell muscle motility, vol 6. Plenum, New York

Wang K, Ramirez-Mitchel R (1983) Ultrastructural morphology and epitope distribution of titin – a giant sarcomere-associated cytoskeletal protein. J Cell Biol 97:986a

Wang K, Wright J (1988) Architecture of the sarcomere matrix of skeletal muscle: immunoelectron microscopic evidence that suggests a set of parallel inextensible nebulin filaments anchored at the Z line. J Cell Biol 107:2199–2212

Wang K, McClure J, Tu A (1979) Titin: major myofibrillar components of striated muscle. Proc Natl Acad Sci USA 76:3698–3702

Wang K, Ramirez-Mitchel R, Palter D (1984) Titin is an extraordinarily long, flexible, and slender myofibrillar protein. Proc Natl Acad Sci USA 81:3685–3689

Wang K, Knipfer M, Huang QQ, Hsu L, van Heerden A, Browning K, Quian X, Stedman H (1990) Structural motifs of nebulin as predicted by sequences of human nebulin cDNA. J Cell Biol 111:428a

Wang K, McCarter R, Wright J, Beverly J, Ramirez-Mitchell R (1993) Viscoelasticity of the sarcomere matrix of skeletal muscles: the titin-myosin composite filament is a dual-stage molecular spring. Biphys J 64:1161–1177

Wright J, Huang Q-Q, Wang K (1993) Nebulin is a full-length template of actin filaments in the skeletal muscle sarcomere: an immunoelectron microscopic study of its orientation and span with site-specific monoclonal antibodies. J Mus Res Cell Motil 14:476–483

Zeviani M, Darras BT, Rizzuto R, Salviati G, Betto R, Bonilla E, Miranda AF, Du J, Samitt C, Dickson G, Walsh FS, Dimauro S, Francke U, Schon EA (1988) Cloning and expression of human nebulin cDNAs and assignment of the gene to chromosome 2q31–q32. Genomics 2:249–256

Tektins from Ciliary and Flagellar Microtubules

R. W. LINCK[1]

1 Introduction

Microtubules are essential eukaryotic cell organelles involved in cell division, cytoplasmic motility, ciliary and flagellar function, and morphogenesis (Alberts et al. 1989; Darnell et al. 1990). Advances in our general understanding of microtubules have come from studies of cilia and flagella, including the discovery of dynein (Gibbons 1963), the basic structure of microtubules (Amos and Klug 1974), and the polarity of microtubule structure and assembly (Allen and Borisy 1974). A more recent discovery concerns the proteins *tektins*.

2 What Are Tektins?

Tektins are a new class of filamentous proteins associated with a stable subset of protofilaments in ciliary and flagellar microtubules. Tektins share some structural homology with intermediate filament proteins but are uniquely different. The following points summarize what is known about these proteins and the specialized portion of the microtubule wall with which they are associated:

1. By extraction with Sarkosyl detergent or by mild heating, ciliary and flagellar microtubules from echinoderms, mollusks, and the protist *Chlamydomonas* can be fractionated into a chemically resistant subset, or ribbon, of 3–4 protofilaments (Fig. 1) (Witman et al. 1972; Meza et al. 1972; Linck 1976; Stephens et al. 1989); we refer to these as *pf-ribbons*.
2. As seen in Fig. 2, the approximate location of the pf-ribbons in axonemal doublet microtubules corresponds to the region of the A-tubule to which the B-tubule and nexin links are attached (Fig. 3) (Linck 1976; Stephens et al. 1989). Indirect evidence also suggests that central pair microtubules may also contain stable pf-ribbons (Steffen and Linck 1988; Linck 1990).
3. Pf-ribbons from sea urchin sperm have a characteristic protein composition (Fig. 4), including: α- and β-tubulin, a pair of 77-/83-kDa polypeptides, and three *tektins*, tektin A (~55-kDa/pI 6.9), tektin B (~51-kDa/pI 6.2), and tektin C

[1] Department of Cell Biology and Neuroanatomy, 4-135 Jackson Hall, 321 Church Street, University of Minnesota, Minneapolis, MN 55455, USA.

45. Colloquium Mosbach 1994
The Cytoskeleton

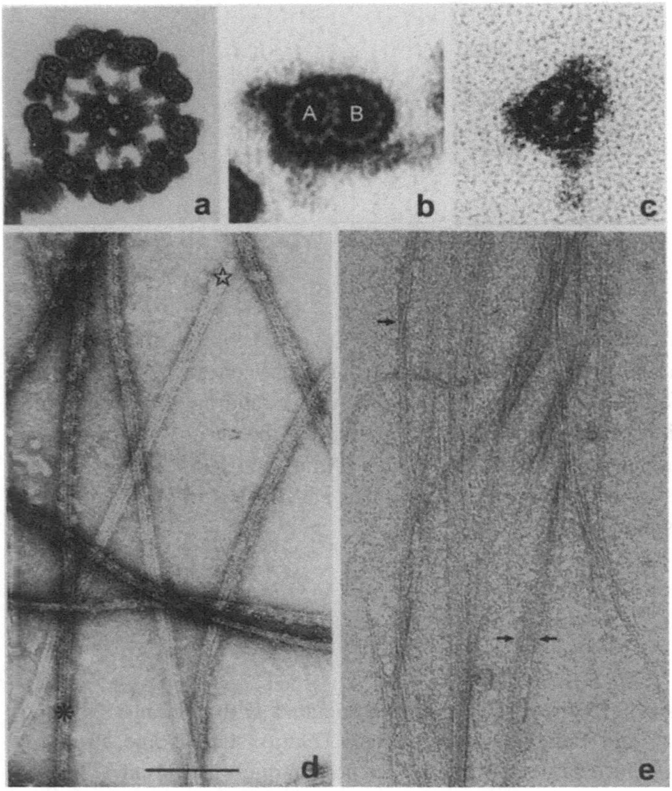

Fig. 1 a–e. Fractionation of sea urchin sperm flagellar axonemes and doublet microtubules into protofilament (pf-) ribbons and tektin filaments, seen by EM. **a** Axoneme. **b** Purified doublet microtubule, showing the A- and B-tubules. **c** A-tubule after solubilization of the B-tubule by thermal fractionation (Stephens 1970; Linck and Langevin 1981), **a–c** Thin sections after tannic acid fixation, showing the protofilaments of the microtubule walls. **d** Ribbons of three protofilaments obtained by extraction of **a, b,** or **c** with 0.5% Sarkosyl detergent; some pf-ribbons (*star*) appear clean, while others (*asterisk*) possess associated material. **e** Tektin filaments obtained by extraction of **a, b, c,** or **d** with 0.5% Sarkosyl + 2 M urea; filaments are highly extended in length and appear with diameters of 2–5 nm (*single arrow*) or in bundles (*between double arrows*). **d** and **e** are negatively stained with uranyl acetate. *Bar* 0.11 μm for **a**, 52 nm for **b**, 35 nm for **c**, 97 nm for **d** and **e**

(~47-kDa/pI 6.15), as well as numerous lower molecular weight polypeptides (Chang and Piperno 1987; Linck et al. 1987).

4. Sea urchin tektins have been characterized biochemically and immunologically. Tektins can be isolated by extraction of pf-ribbons with 0.5% Sarkosyl + 2 M urea, which yields an insoluble residue composed of filaments, 2–5 nm in diameter, as seen by negative stain EM (Fig. 1, Linck and Langevin 1982; Linck et al. 1985; Linck and Stephens 1987). SDS-PAGE analysis indicates that the major components of these preparations are tektins A, B, and C in equimolar amounts (Fig. 4). By amino acid analysis and tryptic peptide mapping tektins A, B, and C

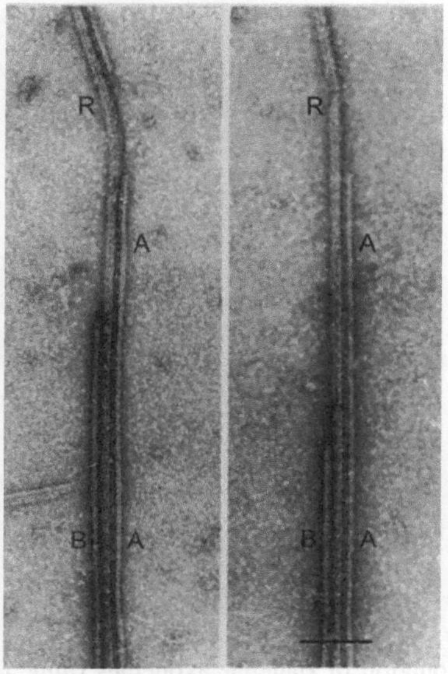

Fig. 2. Negative stain EM showing two doublet microtubules with A- and B-tubules, breaking down into singlet A-tubules, and then into pf-ribbons (*R*). Note that the ribbon extends from the A-tubule. Images such as these suggested the location of the pf-ribbon in the A-tubule of the doublet, as in Fig. 3. *Bar* 50 nm. (Linck 1976)

are 60–70% similar to each other but are unrelated to tubulin (Linck and Stephens 1987). Filaments composed of only tektins A and B are ~70% α-helix (Linck and Langevin 1982), and filaments composed of all three tektins yield strong α-type X-ray patterns (Beese 1984). Monoclonal and polyclonal antibodies against each of the tektins are mainly monospecific for tektins A, B, and C; however, polyclonal antibodies reveal some immunological similarities between tektins A and C under denaturing conditions, and between all tektins under renaturing condition (Chang and Piperno 1987; Linck et al. 1987; Steffen and Linck 1988; Steffen et al. 1994). By two-dimensional isoelectric focusing/SDS-PAGE, tektins B and C each split into several spots, which are recognized by monoclonal antibodies to the respective tektins, indicative of multiple isoforms or posttranslational modifications (Steffen et al. 1994). In addition to the presence of tektins in sea urchin sperm flagella, immunofluorescence microscopy and immunoblotting experiments strongly indicate that tektins are ubiquitous among cilia and flagella, as crossreactions have been observed in ctenophore (coelenterate) comb cilia (Linck et al. 1991), sea urchin (echinoderm) embryonic cilia (Amos et al. 1985; Linck et al.

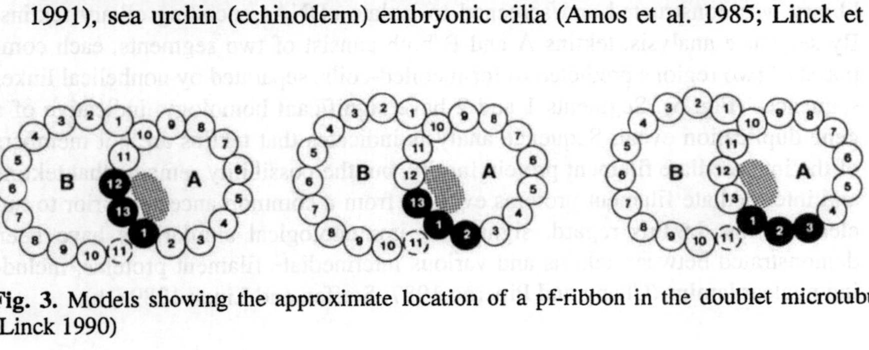

Fig. 3. Models showing the approximate location of a pf-ribbon in the doublet microtubule. (Linck 1990)

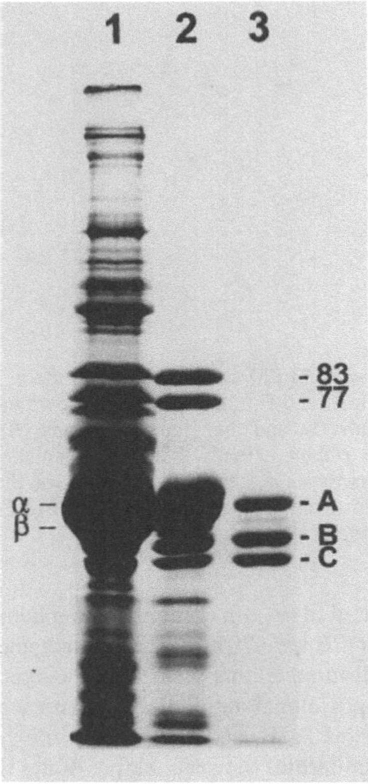

Fig. 4. Fractionation of flagellar axonemes (*lane 1*) into protofilament (pf-) ribbons (*lane 2*), and tektin filaments (*lane 3*), analyzed by SDS-PAGE. Pf-ribbons are composed mainly of a- and β-tubulin, tektins A, B, and C, a 77- and 83-kDa pair of polypeptides, and lower molecular weight polypeptides. Tektin filaments isolated from 0.5% Sarkosyl – 2 M urea are composed of equimolar amounts of tektins A, B, and C, and a small amount of a polypeptide migrating between A and B

1987, Stephens 1989, 1994), scallop (molluskan) gill cilia (Stephens et al. 1989; Steffen and Linck 1992), and bovine (mammalian) tracheal cilia (Hastie et al. 1990).

5. Tektin A and B cDNAs have been sequenced and analyzed, revealing a consensus sequence and a common secondary structure motif (Norrander et al. 1992; Chen et al. 1993). Based on the observation that antibodies against sea urchin sperm flagellar tektins crossreact with sea urchin embryonic cilia, an embryonic expression library was constructed and screened to isolate cDNAs encoding ciliary tektins. By sequence analysis, tektins A and B both consist of two segments, each composed of two regions predicted to form coiled-coils, separated by nonhelical linker sequences (Fig. 5). Segments 1 and 2 have significant homology, indicative of a gene duplication event. Sequence analysis indicates that tektins are not members of the intermediate filament protein family, but the possibility remains that tektins and intermediate filament proteins evolved from a common ancestor, prior to nuclear lamins. In this regard, significant immunological similarities have been demonstrated between tektins and various intermediate filament proteins, including nuclear lamins (Chang and Piperno 1987; Steffen and Linck 1989a, b).

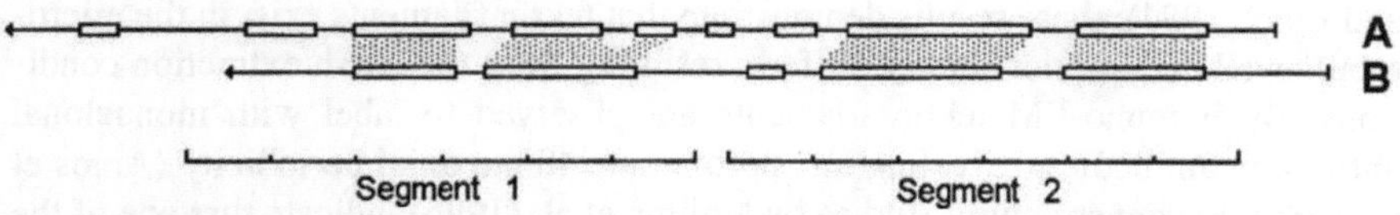

Fig. 5. Model depicting the molecular structure of tektins A and B and the proposed heterodimer. Each polypeptide chain is represented as a linear structure with the N-termini oriented to the left (*arrowheads*). Regions predicted to form α-helix are represented as *open rectangles*, and nonhelical regions as *straight lines*. Regions predicted to form coiled-coil between tektins A and B are *shaded*. Each tektin chain appears to consist of two segments (*1* and *2*) which show sequence homology to each other within a chain and between heterologous chains. If the predicted tektin dimer were a rigid rod as shown, the combined lengths of the two segments would be approximately 48 nm; this length would correspond to 12 tubulin monomers (subdivision) or 6 tubulin dimers along a protofilament. (After Chen et al. 1993)

6. Evidence suggests that tektins form filaments associated with the microtubule wall. By immunofluorescence microscopy all three tektins have been shown to be present in each doublet microtubule (Fig. 6) and possibly the central pair microtubules (Steffen and Linck 1988). By immuno-EM, a residual fibril decorated by anti-tektin antibodies extends from the ends of pf-ribbons (Linck et al. 1985); however, the antibodies do not label tektins along the pf-ribbons, presumably because the tektin epitopes are not exposed to the antibodies. Crosslinking studies show that tektins can be specifically crosslinked within the pf-ribbon, and after Sarkosyl-urea extraction a filament is exposed that is composed of tektins (Pirner

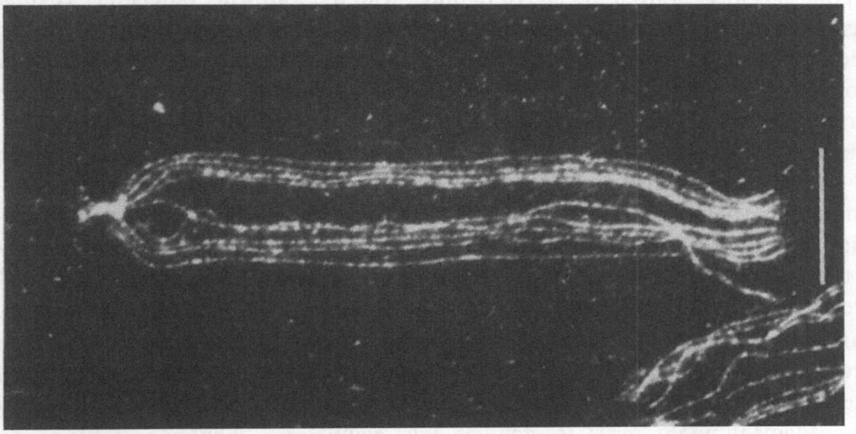

Fig. 6. Immunofluorescence micrograph of *L. pictus* sea urchin sperm after splaying, fixing in –20 °C methanol, and staining with affinity-purified, polyclonal anti-(*L. pictus*)-tektin B. This method results in the splaying of the axonemes into the nine doublet microtubules; the central pair, however, is not preserved in this case. Similar experiments with the other anti-tektin antibodies indicate that all nine doublet microtubules each contain all three tektins A, B, and C. Furthermore, the basal body at the proximal end (*left*) is also intensely stained with anti-tektin B, suggesting the presence of tektins in basal body/centriole triplet microtubules. *Bar* 10 µm. (Steffen and Linck 1988)

and Linck 1994); these results demonstrate that tektin filaments exist in the microtubule wall and are not simply artifacts resulting from the harsh extraction conditions. By immuno-EM tektin filaments are observed to label with monoclonal anti-tektin antibodies, revealing an approximate 48 nm axial periodicity (Amos et al. 1986). Recent structural studies by Nojima et al. (1995) indicate that one of the protofilaments of the pf-ribbons and A-tubules is not composed of tubulin and suggest that this unique protofilament is composed of tektin.

7. By immunofluorescence microscopy, tektins are observed to be associated with basal bodies of sea urchin sperm (Fig. 6) as well as with human centrioles (Steffen and Linck 1988; Steffen et al. 1994). These observations are consistent with tektins being assembled along the A-tubules of ciliary and flagellar doublet microtubules and centriolar triplet microtubules.
8. Finally, monoclonal and polyclonal antibodies against sea urchin tektins have been used to test for the possible presence of tektins outside echinoderm cilia and flagella by immunomicroscopy and immunoblotting techniques. In addition to the crossreactivity seen with cilia, flagella, basal bodies, and centrioles, the anti-tektin antibodies crossreact with molluskan and mammalian centrosomes (Steffen et al. 1994), molluskan and mammalian meiotic/mitotic spindles (Steffen and Linck 1992), and mammalian midbodies (Steffen and Linck 1989b).

3 What Are the Functions of Tektins and Why Might They Be Important?

A hypothesis emerges from the above findings with the following points: (1) Tektins are associated with uniquely stable domains (pf-ribbons) of axonemal and centriolar A-microtubules (Fig. 3) and possibly other kinds of microtubules. (2) Tektins exist as one of the protofilaments in the ribbon. (3) As such, tektin filaments may contribute to forming seam(s) in the microtubule wall.

Stability of the pf-ribbons may be due to tektins (and other proteins) and/or to the local packing of α- and β-tubulin subunits. The arrangement of tubulin dimers in various microtubules has been uncertain, but recent studies of kinesin-decorated microtubules indicate that flagellar A-tubules and cytoplasmic microtubules may all possess the "B-lattice" (Song and Mandelkow 1992), which would generate a "seam" between two of the protofilaments (Linck and Langevin 1981; Mandelkow et al. 1986). Thus, if the A-tubule has a B-type lattice, or even mixed A-type and B-type lattices (cf. Wade et al. 1990), then at least one seam must be present which could create unique binding sites for certain nontubulin proteins. Conceivably, a tektin fibril/protofilament might serve to stabilize two neighboring tubulin protofilaments, giving rise to the chemically unique, stable ribbons of 3-protofilaments (Linck and Langevin 1981).

Although there is as yet no direct evidence for the function of tektins, potentially they could have one or more functions. (1) Tektins could act to stabilize the pf-ribbon and thus the microtubule. The immunological data show a correlation between microtubule stability and the association of tektins or tektin-like proteins. (2) By virtue of their being coiled-coils of repeating subunits, tektin filaments could provide regular assembly/binding sites for certain axonemal components (Figs. 3 and 5), e.g., dynein arms, radial spokes, nexin links, B-tubules, and centriolar C-tubules (Norrander et al. 1992). All these components index on a fundamental 96-nm axial repeat that is a multiple of the underlying tubulin lattice (Amos and Klug 1974; Amos et al. 1976), but tubulin alone does not appear to have the capacity for such complex positional information. Interestingly, pf-ribbons extracted from ciliary axonemes are retained in a nine-fold array, presumably by the nexin filament system attached to the pf-ribbons of the A-tubules (Stephens et al. 1989). (3) Lastly, coiled-coil filaments in the microtubule wall would obviously affect the elastic bending properties of axonemal microtubules, important to the regulation of propagated bending waves (Brokaw 1991; Lindemann et al. 1992).

With this work as a foundation, we hope to elucidate the function of tektins by continuing to investigate the structure and assembly of tektins and tubulin in sea urchin sperm flagella, and to elucidate tektin function in vivo, using the genetically manipulated organism *Chlamydomonas*.

Acknowledgments. I wish to thank Eckhard Mandelkow for the opportunity to present this work at the Mosbacher Colloquium. I also thank Mark Pirner for his assistance in preparing several of the figures. This research was supported by USPHS grant GM35648 and NSF-RTG grant DIR-9113444.

References

Alberts B, Bray D, Lewis J, Raff M, Roberts K & Watson JD (1989) Molecular biology of the cell. Garland Publishing, Inc, NY

Allen C & Borisy G (1974) Structural polarity and directional growth of microtubules of *Chlamydomonas* flagella. J Mol Biol 90:381–402

Amos WB, Amos LA and Linck RW (1985) Proteins similar to flagellar tektins are detected in cilia but not in cytoplasmic microtubules. Cell Motil 5:239–250

Amos WB, Amos LA & Linck RW (1986) Studies of tektin filaments from flagellar microtubules by immunoelectron microscopy. J Cell Sci Suppl 5:55–68

Amos LA & Klug A (1974) Arrangement of subunits in flagellar microtubules. J Cell Sci 14:523–549

Amos LA, Linck RW & Klug A (1976) Molecular structure of flagellar microtubules. In: Cell motility (eds R Goldman, T Pollard & J Rosenbaum), pp 847–867. New York: Cold Spring Harbor Laboratory

Beese L (1984) Microtubule structure: an analysis by X-ray diffraction. PhD thesis. Brandeis University, Waltham, MA

Brokaw CJ (1991) Microtubule sliding in swimming sperm flagella: direct and indirect measurements on sea urchin and tunicate spermatozoa. J Cell Biol 114:1201–1215

Chang XJ & Piperno G (1987) Cross-reactivity of antibodies for flagellar tektins and intermediate filament subunits. J Cell Biol 104:1563–1568

Chen R, Perrone CA, Amos LA & Linck RW (1993) Tektin B1 from ciliary microtubules: primary structure as deduced from the cDNA sequence and comparison with tektin A1. J Cell Sci 106:909–918

Darnell J, Lodish H & Baltimore D (1990) Molecular cell biology. WH Freeman, Inc, NY

Gibbons IR (1963) Studies on the protein components of cilia from *Tetrahymena pyriformis*. Proc Natl Acad Sci USA 50:1002–1010

Hastie AT, Krantz MJ, Colizzo FP & Evans LP (1990) Investigation of cross-reactivity of bovine tracheal epithelial cytokeratins and ciliary tektins. J Cell Biol 111:40a
Linck RW (1976) Flagellar doublet microtubules: fractionation of minor components and α-tubulin from specific regions of the A-tubule. J Cell Sci 20:405–439
Linck RW (1990) Tektins and microtubules. Adv Cell Biol 3:35–63
Linck RW, Amos LA & Amos WB (1985) Localization of tektin filaments in microtubules of sea urchin sperm flagella by immunoelectron microscopy. J Cell Biol 100:126–135
Linck RW, Goggin MJ, Norrander JM & Steffen W (1987) Characterization of antibodies as probes for structural and biochemical studies of tektins from ciliary and flagellar microtubules. J Cell Sci 88:453–466
Linck RW & Langevin GL (1981) Reassembly of flagellar B (αβ) tubulin into singlet microtubules: consequences for cytoplasmic microtubule structure and assembly. J Cell Biol 89:323–337
Linck RW & Langevin GL (1982) Structure and chemical composition of insoluble filamentous components of sperm flagellar microtubules. J Cell Sci 58:1–22
Linck RW & Stephens RE (1987) Biochemical characterization of tektins from sperm flagellar doublet microtubules. J Cell Biol 104:1069–1075
Linck RW, Stephens RE & Tamm SL (1991) Evidence for tektins in cilia from the ctenophore, *Mnemiopsis leidyi*. In: Comparative spermatology 20 years after. B Baccetti, ed Serono Symposia Publications, Vol 75, Raven Press, New York, pp 391–395
Lindemann CB, Orlando A & Kanous KS (1992) The flagellar beat of rat sperm is organized by the interaction of two functionally distinct populations of dynein bridges with a stable central axonemal partition. Cell Motil Cytoskel 102:249–260
Mandelkow EM, Schultheiss R, Rapp R, Müller M & Mandelkow E (1986) On the surface lattice of microtubules: helix starts, protofilament number, seam, and handedness. J Cell Biol 102:1067–1073
Meza I, Huang B & Bryan J (1972) Chemical heterogeneity of protofilaments forming the outer doublets from sea urchin flagella. Exp Cell Res 74:535–540
Nojima D, Linck RW & Egelman EH (1995) At least one of the protofilaments in flagellar microtubules is not composed of tubulin. Cur Biol 5: In press
Norrander JM, Amos LA & Linck RW (1992) Primary structure of tektin A1: comparison with intermediate-filament proteins and a model for its association with tubulin. Proc Natl Acad Sci USA 89:8567–8571
Pirner MA & Linck RW (1994) Tektins are heterodimeric polymers in flagellar microtubules with axial periodicities matching the tubulin lattice. J Biol Chem 269:31800–31806
Song YH & Mandelkow E (1993) Recombinant kinesin motor domain binds to beta-tubulin and decorates microtubules with a B-surface lattice. Proc Natl Acad Sci USA 90:1671–1675
Steffen W & Linck RW (1988) Evidence for tektin-related components in axonemal microtubules and centrioles. Proc Natl Acad Sci USA 85:2643–2647
Steffen W & Linck RW (1989a) Relationship between tektins and intermediate filament proteins: an immunological study. Cell Motil Cytoskel 14:359–371
Steffen W & Linck RW (1989b) Tektins in ciliary and flagellar microtubules and their association with other cytoskeletal systems. In: Cell movement, Vol 2: Kinesin, dynein and microtubule dynamics, Chapter 2. FD Warner and JR McIntosh, eds, Alan R Liss, Inc New York, p 67–81
Steffen W & Linck RW (1992) Evidence for a non-tubulin spindle matrix and for spindle components immunologically related to tektin filaments. J Cell Sci 101:809–822
Steffen W, Fajer EA & Linck RW (1994) Centrosomal components immunologically related to tektins from ciliary and flagellar microtubules. J Cell Sci. In press
Stephens RE (1970) Thermal fractionation of outer doublet microtubules into A- and B-subfiber components: A- and B-tubulin. J Mol Biol 47:353–363
Stephens RE (1989) Quantal synthesis and ciliary length in sea urchin embryos. J Cell Sci 92:403–413
Stephens RE (1994) Tubulin and tektin in sea urchin embryonic cilia: pathways of protein incorporation during turnover and regeneration. J Cell Sci 107:683–692

Stephens RE, Oleszko-Szuts S & Linck RW (1989) Retention of ciliary ninefold structure after removal of microtubules. J Cell Sci 92:391–402

Wade RH, Crétien D & Job D (1990) Characterization of microtubule protofilament numbers. How does the surface lattice accommodate? J Mol Biol 212:775–786

Witman GB, Carlson K, Berliner J & Rosenbaum JL (1972) *Chlamydomonas* flagella. I Isolation and electrophoretic analysis of microtubules, matrix, membranes, and mastigonemes. J Cell Biol 54:507–539

Actin and Actin-Binding Proteins in the Motility of *Dictyostelium*

A. A. NOEGEL[1], B. KÖPPEL[1], U. GOTTWALD[1], W. WITKE[1], R. ALBRECHT[1], and M. SCHLEICHER[2]

1 Introduction

Dictyostelium discoideum cells are highly motile throughout all stages of development and directed cell motility is essential for morphogenesis to occur (Williams and Jermyn 1991). Cells move by pseudopod extension, and formation and retraction of pseudopods in response to a chemotactic stimulus is critical for directed migration. Folate acts as chemotactic agent during growth, cAMP is the active compound during development. The understanding of chemotactic signaling has advanced considerably (Devreotes 1989), but its linkage to intracellular changes that lead to cell motility is less clear (reviewed in Schleicher and Noegel 1992). Early observations indicated that chemoattractant-induced pseudopod formation correlates temporally and spatially with the polymerization of actin, and peaks of F-actin occur by 15–60 s after stimulation (McRobbie and Newell 1983; Hall et al. 1988). Five different steps in actin changes can be distinguished (Fig. 1), and these steps were found to coincide with retraction of the existing pseudopod and generation of new pseudopods (Dhamawardhane et al. 1989). Parallel processes are an influx of Ca^{2+} and an efflux of protons leading to an alkalinization of the cytoplasm. The phosphoinositides are also affected with an increase of IP_3 as fast as 6 s after stimulation at the expense of PIP and PIP_2 (Europe-Finner et al. 1991).

2 Actin-Binding Proteins and Actin Assembly

F-actin assembly and disassembly can be regulated by ionic conditions and pH. As key regulators, however, actin-binding proteins are thought to affect actin polymerization and the three-dimensional assembly of F-actin in vivo, and the action of distinct proteins is proposed to be responsible for different steps in the cycle (Condeelis 1993a, b; Stossel 1993). In the initial step of depolymerization during the disintegration of the existing pseudopod, the activities of F-actin fragmenting proteins and G-actin-sequestering proteins are required, whereas for the following polymerization steps, actin nucleating proteins provide actin nuclei and actin crosslinking proteins stabilize the network. Proteins exhibiting such activities have been isolated and their characterization has been extended in several cases even to the atomic level

[1] Max-Planck-Institut für Biochemie, 82152 Martinsried FRG, Institut für Zellbiologie.
[2] Ludwig-Maximilians-Universität, 80336 Munich, FRG.

45. Colloquium Mosbach 1994
The Cytoskeleton

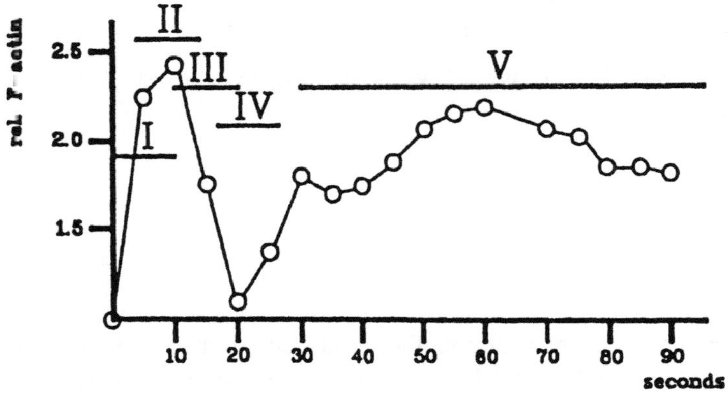

Fig. 1. Time course of actin polymerization after stimulation with chemoattractant. *Phase I* 1–10 s after stimulation. Binding of chemoattractant leads to a hold of the cell and disintegration of the leading pseudopod. An influx of Ca^{2+} takes place, efflux of protons leads to the alkalinization of the cytoplasm, the polyphosphoinositides are affected in their relative concentrations. *Phase II* 5–15 s after stimulation. A creation of new actin filaments occurs and a doubling of filament number is observed. *Phase III* 10–20 s after stimulation. After the initial peak of F-actin a rapid depolymerization takes place. *Phase IV* 15–35 s after stimulation. The cringe reaction, a rounding-up of the cell, is the result of a myosin II based concentration of the cortical actin meshwork. *Phase V* 30–50 s after stimulation. A second and long-lasting filament formation takes place coinciding with the formation of new pseudopods

(Habazettl et al. 1992; MacLaughlin et al. 1993; Schutt et al. 1993; Vinson et al. 1993). The ease of cultivation of *Dictyostelium* cells and their availability in large quantities has facilitated biochemical studies of actin-binding proteins in *D. discoideum*. Their isolation and characterization is therefore well advanced and proteins corresponding to the ones known from higher eukaryotes as well as novel proteins have been identified (Fig. 2) (reviewed in Luna and Condeelis 1991; Schleicher and Noegel 1992). Furthermore, *D. discoideum* is amenable to genetic manipulation and thus appears to be well suited for dissection of the molecular reactions occuring during chemotaxis in combination with a biochemical analysis of the components (Noegel et al. 1989).

2.1 Actin-Binding Proteins from D. discoideum

2.1.1 G-actin Binding Proteins

Two profilin isoforms are present in *D. discoideum*. They differ slightly in their affinity for F-actin (5×10^{-6} M for profilin I, 1.8×10^{-6} for profilin II) and also have a different expression pattern during development, with profilin II being more prominent during growth and early aggregation, whereas profilin I accumulates maximally in the late aggregation and culmination stage (Haugwitz et al. 1991). Profilin I and II are 55% identical at the amino acid level and 62% identical at the DNA level. Several roles have been proposed for profilins (for review see Theriot and Mitchison 1993).

Initially, they were characterized as actin-sequestering proteins but this activity has been challenged by more recent findings, which point out that profilin can also promote actin polymerization by exchanging ADP for ATP in actin and therefore allowing polymerization to occur. By a third mechanism, profilin promotes polymerization by lowering the critical concentration (Pantaloni and Carlier 1993). Furthermore, the binding of profilin to PIP_2 has opened a different avenue of research and provided a hint that profilin may also be involved in signal transduction (Machesky and Pollard 1993).

2.1.2 F-Actin Capping Proteins

Capping proteins inhibit actin polymerization by binding to the fast-growing end of a filament. In *D. discoideum* cap32/34 has been isolated which belongs to the group of heterodimeric capping proteins (Hartmann et al. 1989). Its capping activity is inhibited by phospholipids (Haus et al. 1991). After chemotactic stimulation this protein is found associated with the Triton-insoluble cytoskeleton during the initial actin polymerization step (step II) (Brink 1989). This observation renders this protein a very good candidate for participating in the early events after stimulation since such an activity appears to be imperative for chemotaxis to occur. Cap32/34 when isolated from *D. discoideum* can be associated with a 70-kDa protein which was identified as a heat shock cognate protein (hsc70) both by protein and cDNA sequencing. Its interaction with cap32/34 was further studied with purified components separately expressed in *E. coli*. In this way, two activities of hsc70 could be distinguished. For one, the N-terminal region of hsc70 exhibiting an actin-like ATPase domain was found to enhance the capping activity of cap32/34, whereas the more hydrophobic C-terminal domain helped in proper folding of the protein (Haus et al. 1993). These findings add a further level of complexity to the regulation of actin assembly.

2.1.3 F-Actin Severing Proteins

Severin, a 40000-Da protein from *D. discoideum* is a Ca^{2+}-regulated F-actin fragmenting protein. It is a member of a group of structurally and functionally highly homologous proteins composed of three (severin, fragmin) or six (gelsolin, villin) homologous repeats (Janmey 1993). The functions of these repeats have been thoroughly analyzed and models of how these proteins work have been put forward (Pope et al. 1991; Eichinger et al. 1991; Eichinger and Schleicher 1992). Furthermore, the structures of domain 1 of gelsolin (McLaughlin et al. 1993) and domain 2 of severin (Holak and Schleicher, unpubl.) have been solved and found to exhibit the same fold.

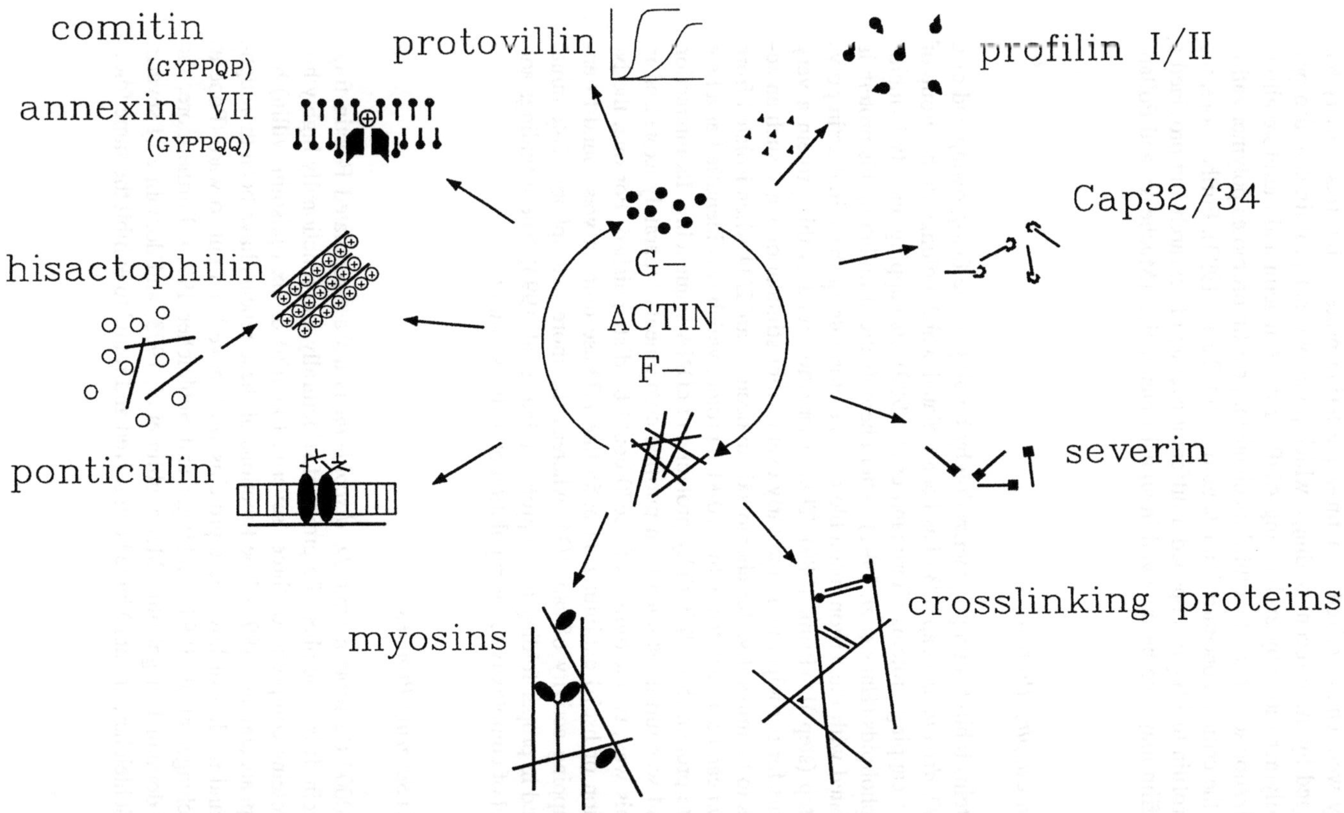

Fig. 2. Actin-binding proteins in *D. discoideum.* Actin-binding proteins with capping (*protovillin*: Hofmann et al. 1992, 1993; *cap32/34*: Hartmann et al. 1989; Haus et al. 1992, 1993) or severing functions (*severin*: André et al. 1988) lower the average length of actin filaments. Crosslinking proteins (*α-actinin, gelation factor:* Noegel et al. 1989) connect actin filaments and stabilize the network. *Myosins* in conjunction with filaments produce contractile forces (Spudich 1989), and membrane-associated proteins establish an interaction of the microfilament system with the plasma membrane (*ponticulin*: Wuestehube and Luna 1989) or intracellular membranes (*comitin*: Weiner et al. 1993)

Table 1. Mutants in actin-binding proteins in *D. discoideum*. Selected actin-binding proteins have been isolated, characterized and, where indicated, the genes have been inactivated

Proteins	App. MW	Purification	Antibodies	Cloning	Sequencing	Gene disruption	Transformants	Expression in E. coli	Lipid binding	NMR studies
Gelation factor	120	+	+	+	+	+	+	+		
α-Actinin	97	+	+	+	+	+	+	+	?	
Severin	40	+	+	+	+	+	+	+	+	+
Cap32/34	32/34	+	++	++	++		++	++	+	
Hsc70	70	+	+	+	+			+		
Profilin I/II	12	++	++	++	++	+	+	+	?	
Hisactophilin I/II	17	++	+	++	++	++		+	+	+
Comitin	24	+	+	+	+		+	+		
Protovillin	100	+	+	+	+	+		+	+	

2.1.4 F-Actin Crosslinking Proteins

Several F-actin crosslinking proteins are present in *D. discoideum*. Most prominent are α-actinin and 120-kDa gelation factor. Both are rod-shaped molecules consisting of two identical subunits which are arranged in an antiparallel fashion. The monomers have an actin-binding site located at the N-terminus (Noegel et al. 1987; Noegel et al. 1989). This actin-binding site is highly homologous among a-actinin and gelation factor and is also found in spectrin, filamin, dystrophin, and fimbrin. In the dimer the actin binding sites are located at both ends of the molecules. Actin crosslinking of α-actinin is inhibited by micromolar concentrations of Ca^{2+}, the regulation of the gelation factor is not known.

3 Motility in *Dictyostelium* Mutants Defective in Actin-Binding Proteins

For the analysis of the role of actin binding proteins during chemotactic movement, *D. discoideum* seems to be especially well suited because it is a haploid organism allowing the convenient isolation of mutants. For mutant generation, both chemical mutagenesis as well as targeted gene inactivation using homologous recombination can be applied. These approaches led to the isolation of mutants lacking such key regulators as the F-actin fragmenting protein severin (André et al. 1989), the F-actin network stabilizing proteins α-actinin (Wallraff et al. 1986; Witke et al. 1987; Schleicher et al. 1988) and gelation factor (Brink et al. 1990; Cox et al. 1992) as well as the G-actin sequestering proteins profilin I and II (Table 1).

Severin and the two crosslinking proteins α-actinin and gelation factor exhibit activities which appear to be essential for pseudopod retraction and extension. In mutants lacking these proteins however no significant impairment of motility was noted (Table 2). With α-actinin and the gelation factor being modular isoforms in a next

Table 2. Motility of AX2 wild-type and mutant cells. In GHR the gelation factor gene, in AHR the α-actinin gene, in DSC211 cells the severin gene has been inactivated, AGHR2 and GA1 lack α-actinin and gelation factor, and profI/II-minus is deficient for profilin I and profilin II. The determination of the speed was done according to Fisher et al. (1989). n is the number of independent experiments The standard deviation is given

Strain	n	Speed (μm/min) – cAMP		Speed (μm/min) + cAMP	
AX2	6	6.9	(+/–1.7)	10.2	(+/–1.0)
GHR	5	6.1	(+/–1.1)	8.3	(+/–0.8)
AHR	5	9.5	(+/–2.5)	10.7	(+/–2.0)
AGHR2	7	4.8	(+/–0.8)	5.5	(+/–0.3)
GA1	2	3.9	(+/–0.9)	7.9	(+/–2.9)
pI/II-minus	4	4.1	(+/–1.0)	6.4	(+/–1.1)
DSC211	6	8.31	(+/–1.78)	13.8	(+/–2.12)

step both proteins were inactivated. This double deficiency had marked effects on motility and chemotactic movement. Motility was also impaired in mutants lacking profilin I and II, whereas a deficiency in either one of the two proteins had no observable effect (Table 2).

4 Morphogenesis in *Dictyostelium* Mutants Lacking Actin-Binding Proteins

D. discoideum cells, when starved, undergo a developmental cycle. Single cells form an aggregate and differentiate into cell types within this aggregate. Ordered movement of the cells leads to the formation of a fruiting body consisting of a basal plate, a stalk, and a spore head. Cell motility therefore is an essential aspect of morphogenesis. In mutants with defects in a single actin-binding protein development was nearly normal; however, in mutants GA1 and AGHR2 lacking the two crosslinking proteins α-actinin and gelation factor (Witke et al. 1992) and in a mutant lacking both profilins (ProfI/II$^-$), no fruiting body formation was observed. In AGHR2 development stopped after aggregation (Fig. 3). The early aggregation markers cAMP receptor cAR1 (Fig. 4) and cell adhesion molecule csA were still present whereas cell type specific markers like the products of the ecmA, ecmB, or D19 gene were not detectable. In contrast, in the ProfI/II$^-$ mutant differentiation into cell types took place and the aggregates were able to culminate. Further morphogenesis, however, did not take place. In both cases the observed defects could be reversed by introducing a func-

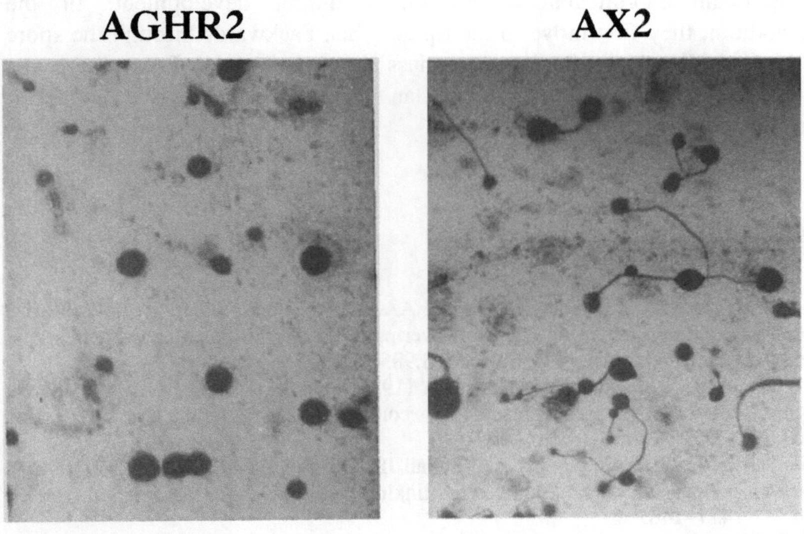

Fig. 3. Development of strain AGHR2 lacking α-actinin and gelation factor. AGHR2 and for control AX2 wild-type cells were plated onto phosphate agar and allowed to develop. The photographs were taken after 24 h. *AX2* cells had formed fruiting bodies, *AGHR2* cells had aggregated. No further development took place in this strain

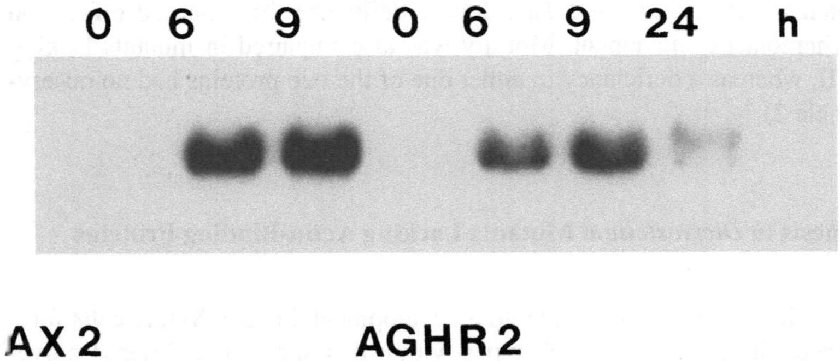

Fig. 4. Expression of a development specific marker in AGHR2 cells. AGHR2 and AX2 cells were allowed to develop in suspension culture at a density of 1 x 10^7 cells per ml. Aliquots were taken at the indicated time points and total RNA was isolated. Ten µg of RNA were loaded onto a 1.2% agarose gel and the RNA separated in the presence of 6% formaldehyde. The resulting Northern blot was probed with cAR1-specific cDNA encoding the cAMP receptor I (Klein et al. 1988). The RNA was detectable in both strains at the same developmental time point. In AGHR2 the cAR1-specific mRNA was present in slightly lower amounts compared to wild type

tional crosslinker or profilin gene, respectively. Morphogenesis also took place when mutant cells were mixed with wild-type cells. Using wild-type cells which expressed β-galactosidase, the fate of these cells was directly followed in mixed aggregates with GA1 or AGHR2. In these synergy experiments wild-type cells formed mostly stalk cells whereas the mutant cells differentiated into spores. Presumptive stalk cells undergo significant morphogenetic movements during development. In the pseudoplasmodium, they first move to the tip and then backwards through the spore mass to form the stalk which lifts the spore mass above the ground. From the results obtained with the mutants, it is evident that an intact cytoskeleton is essential for morphogenesis to occur.

References

André E, Brink M, Gerisch G, Isenberg G, Noegel AA, Schleicher M, Segall JE & Wallraff E (1989) A *Dictyostelium* mutant deficient in severin, an F-actin fragmenting protein, shows normal motility and chemotaxis. J Cell Biol 108:985–995

André E, Lottspeich F, Schleicher M & Noegel A (1988) Severin, gelsolin and villin share a homologous sequence in regions presumed to contain F-actin severing domains. J Biol Chem 263:722–727

Brink M, Gerisch G, Isenberg G, Noegel AA, Segall JE, Wallraff E & Schleicher M (1990) A *Dictyostelium* mutant lacking an F-actin cross-linking protein, the 120 kD gelation factor. J Cell Biol 111:1477–1489

Condeelis J (1993a) Life at the leading edge: the formation of cell protrusions. Annu Rev Cell Biol 9:411–444

Condeelis J (1993b) Understanding the cortex of crawling cells: insights from *Dictyostelium*. Trends in Cell Biol 3:371–376

Cox D, Condeelis J, Wessels D, Soll D, Kern H & Knecht DA (1992) Targeted disruption of the ABP-120 gene leads to cells with altered motility. J Cell Biol 116:943–955

Dharmawardhane S, Warren V, Hall AL & Condeelis J (1989) Changes in association of actin-binding proteins with the actin cytoskeleton during chemotactic stimulation of *Dictyostelium discoideum*. Cell Motil Cytoskel 13:57–63

Devreotes P (1989) Cell-cell interactions in *Dictyostelium* development. TIG 5:242–245

Eichinger L, Noegel AA & Schleicher M (1991) Domain structure in actin-binding proteins: expression and functional characterization of truncated severin. J Cell Biol 112:665–676

Eichinger L & Schleicher M (1992) Characterization of actin- and lipid-binding domains in severin, a Ca^{2+}-dependent F-actin fragmenting protein. Biochemistry 31:4779–4787

Europe-Finner GN, Gammon B & Newell PC (1991) Accumulation of (^{3}H)-inositol into inositol polyphosphates during development of *Dictyostelium*. Biochem Biophys Res Commun 130:1115–1122

Fisher PR, Merkl R & Gerisch G (1989) Quantitative analysis of cell motility and chemotaxis in *Dictyostelium discoideum* by using an image processing system and a novel chemotaxis chamber providing stationary chemical gradients. J Cell Biol 108:973–984

Habazettl J, Gondol D, Wiltscheck R, Otlewski J, Schleicher M & Holak TA (1992) Structure of hisactophilin is similar to interleukin-1β and fibroblast growth factor. Nature 359:855–858

Hall AL, Schlein A & Condeelis J (1988) Relationship of pseudopod extension to chemotactic hormone-induced actin polymerization in amoeboid cells. J Cell Biochem 37:285–299

Hartmann H, Noegel AA, Eckerskorn C, Rapp S & Schleicher M (1989) Ca^{2+}independent F-actin capping proteins. J Biol Chem 264:12639–12647

Haugwitz M, Noegel AA, Rieger D, Lottspeich F & Schleicher M (1991) *Dictyostelium discoideum* contains two profilin isoforms that differ in structure and function. J Cell Sci 100:481–489

Haus U, Hartmann H, Trommler P, Noegel AA & Schleicher M (1991) F-actin capping by cap32/34 requires heterodimeric conformation and can be inhibited with PIP_2. Biochem Biophys Res Commun 181:883–839

Haus U, Trommler P, Fisher PR, Hartmann H, Lottspeich F, Noegel AA & Schleicher M (1993) The heat shock cognate protein from *Dictyostelium* affects actin polymerization through interaction with the actin-binding protein cap32/34. EMBO J 12:3763–3771

Hofmann A, Eichinger L, André E, Rieger D & Schleicher M (1992) Cap 100, a new PIP_2 regulated protein that caps actin filaments but does not nucleate actin-assembly. Cell Motil Cytoskel 23:133–144

Hofmann A, Noegel AA, Bomblies L, Lottspeich F & Schleicher M (1993) The 100 kDa F-actin capping protein of *Dictyostelium* amoebae is a villin prototype ("protovillin"). FEBS Lett 328:71–76

Janmey P (1993) A slice of the actin. Nature 364:675–676

Luna EJ & Condeelis J (1990) Actin-associated proteins in *Dictyostelium discoideum*. Dev Genet 11:328–322

Klein PS, Sun TJ, Saxe III CL, Kimmel AR, Johnson RL & Devreotes PN (1988) A chemoattractant receptor controls development in *Dictyostelium discoideum*. Science 241:1467–1472

Machesky LM & Pollard TD (1993) Profilin as a potential mediator of membrane-cytoskeleton communication. Trends in Cell Biol 3:381–385

McLaughlin PJ, Gooch JT, Mannherz HG & Weeds AG (1993) Structure of gelsolin segment 1-actin complex and the mechanism of filament severing. Nature 364:685–692

McRobbie SJ & Newell PC (1983) Changes in actin associated with the cytoskeleton following chemotactic stimulation of *Dictyostelium discoideum*. Biochem Biophys Res Commun 115:351–359

Noegel A, Witke W & Schleicher M (1987) Calcium-sensitive non-muscle α-actinin contains EF-hand structures and highly conserved regions. FEBS Lett 221:391–396

Noegel AA, Leiting B, Witke W, Gurniak C, Harloff H, Hartmann H, Wiesmüller E & Schleicher M (1989) Biological roles of actin-binding proteins in *Dictyostelium discoideum* examined using genetic techniques. Cell Motil Cytoskel 14:69–74

Noegel AA, Rapp S, Lottspeich F, Schleicher M & Stewart M (1989) The *Dictyostelium* gelation factor shares a putative actin binding site with a-actinins and dystrophin and also has a rod domain containing six 100-residue motifs that appear to have cross-beta conformation. J Cell Biol 109:607–618

Pantaloni D & Carlier MF (1993) How profilin promotes actin filament assembly in the presence of thymosin-β_4. Cell 75:1007–1014

Pope B, Way M & Weeds AG (1991) Two of the three actin-binding domains of gelsolin bind to the same subdomain of actin. FEBS Lett 280:70–74

Schleicher M, Noegel A, Schwarz T, Wallraff E, Brink M, Faix J, Gerisch G & Isenberg G (1988) A *Dictyostelium* mutant with severe defects in α-actinin: its characterization using cDNA probes and monoclonal antibodies. J Cell Sci 90:59–71

Schleicher M & Noegel AA (1992) Dynamics of the *Dictyostelium* cytoskeleton during chemotaxis. The New Biol 4:461–472

Schutt CE, Myslik JC, Royzycki MD, Goonesekere NCW & Lindberg U (1993) The structure of crystalline profilin-β-actin. Nature 365:810–816

Spudich JA (1989) In pursuit of myosin function. Cell Regul 1:1–11

Stossel TP (1993) On the crawling of animal cells. Science 260:1086–1094

Theriot JA & Mitchison TJ (1993) The three faces of profilin. Cell 75:835–838

Vinson V, Archer S, Lattman E, Pollard T & Torchia D (1993) Three-dimensional solution structure of *Acanthamoeba* profilin-I. J Cell Biol 122:1277–1283

Wallraff E, Schleicher M, Modersitzki M, Rieger D, Isenberg G & Gerisch G (1986) Selection of *Dictyostelium* mutants defective in cytoskeletal proteins: use of an antibody that binds to the ends of α-actinin rods. EMBO J 5:61–67

Weiner OH, Murphy J, Griffiths G, Schleicher M & Noegel AA (1993) The actin-binding protein comitin (p24) is a component of the Golgi apparatus. J Cell Biol 123:23–34

Williams JG & Jermyn KA (1991) Cell sorting and positional differentiation during *Dictyostelium* morphogenesis. In: Cell-cell-interactions in early development, Wiley-Liss, Inc pp 261–272

Witke W, Nellen W & Noegel A (1987) Homologous recombination in the *Dictyostelium* α-actinin gene leads to an altered mRNA and lack of the protein. EMBO J 6:4143–4148

Witke W, Schleicher M & Noegel AA (1992) Redundancy in the microfilament system: abnormal development of *Dictyostelium* cells lacking two F-actin cross-linking proteins. Cell 68:53–62

Wuestehube LJ & Luna EJ (1987) F-actin binds to the cytoplasmic surface of ponticulin, a 17-kD integral glycoprotein from *Dictyostelium discoideum* plasma membranes. J Cell Biol 105:1741–1751

Analysis of Cell Motility in Living Cells

J. M. SANGER[1] and J. W. SANGER[1]

1 Introduction: Examples of Motile Events in Living Cells

We have been studying how cell divide by employing a variety of optical methods to determine what cytoskeletal elements are used and how their position is determined. Cell division consists of two main processes: (1) mitosis, the separation of the daughter chromosomes to the distal ends of the mother cell, and (2) cytokinesis, the contraction of the area between the two sets of separating chromosomes to form two new daughter cells (Fig. 1; Sanger et al. 1994). These two processes involve two different cytoskeletal systems, i.e., microtubules and microtubule motors, and actin and myosin motors. Phase-contrast microscopy allows a clear view of the two motile processes (Fig. 2; Sanger et al. 1989). Numerous studies have indicated that the plane of the cytokinesis is determined by the position of the mitotic spindle in the cell (see review by Rappaport 1986). Whereas phase-contrast microscopy allows the visualization of the chromosomes and sometimes the vesicular components of the cell, fluorescence microscopy and fluorescent dyes allow the visualization of individual cytoskeletal and membranous components of the living cells. The microinjection of fluorescently labeled actin and myosin probes allowed us to follow the timing of the deposition of actin and myosin in the cleavage furrows (Fig. 3; Sanger et al. 1989a, 1994), showing that actin and myosin begin to accumulate at the cell surfaces above the former metaphase plate at mid-anaphase. If the mitotic spindle is displaced to one side of the cell, the cleavage furrow proteins assemble on the same side (Fig. 3). Our working hypothesis is that the interzonal microtubules that emanate from opposite centrosomes must interact with the cell surfaces to allow actin and myosin cytoskeleton units to localize and assemble contractile units. The development of fluorescent probes such as the fluorescent dye DiOC6(3) (i.e., 3.3'-dihexyloxacarbocyanine iodide) for membranes (see review by Terasaki 1989) has allowed us and others to follow the distribution of vesicles inside and outside the mitotic spindle (Sanger et al. 1989b). Confocal microscopy can be coupled to these fluorescent methods in living cells to localize the probes to a narrow focal plane (Fig. 4; Waterman-Storer et al. 1994).

We have used these microscopic methods to study other processes in living cells, e.g., myofibrillogenesis (Sanger et al. 1986; Rhee et al. 1994) or bacterial infections (Sanger et al. 1992). Fluorescence microscopy has enabled us to detect the linear precursors of mature myofibrils in both skeletal and cardiac muscle cells (Sanger et al.

[1] Department of Cell and Developmental Biology, University of Pennsylvania School of Medicine, Philadelphia, PA 19104–6058, USA.

45. Colloquium Mosbach 1994
The Cytoskeleton

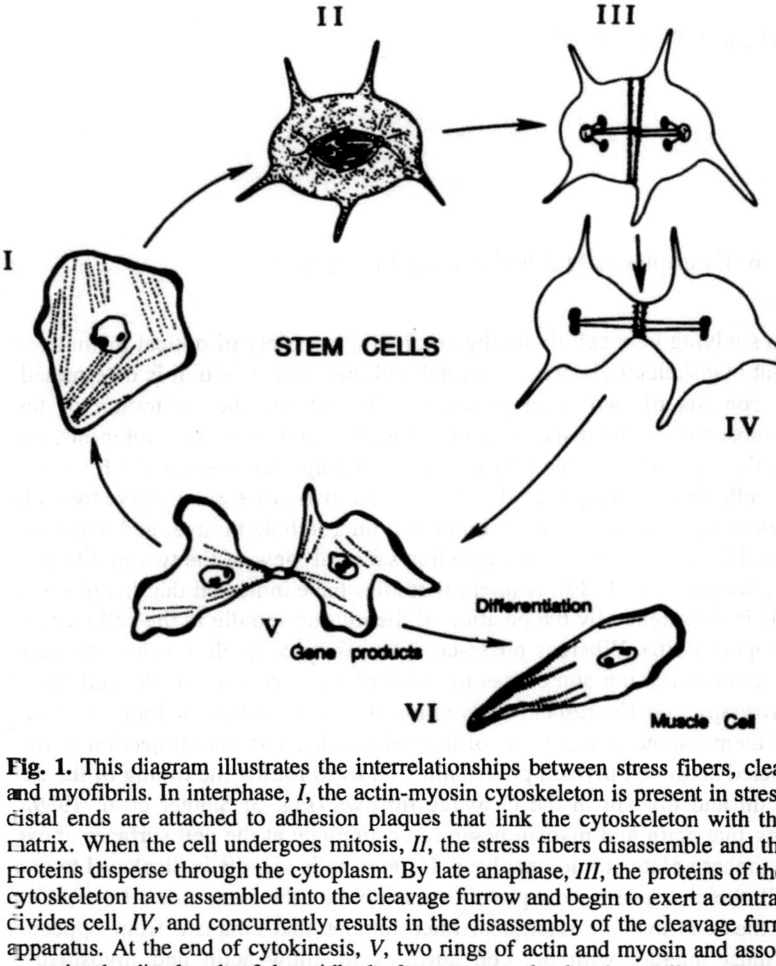

Fig. 1. This diagram illustrates the interrelationships between stress fibers, cleavage furrows, and myofibrils. In interphase, *I*, the actin-myosin cytoskeleton is present in stress fibers whose distal ends are attached to adhesion plaques that link the cytoskeleton with the extracellular matrix. When the cell undergoes mitosis, *II*, the stress fibers disassemble and their constituent proteins disperse through the cytoplasm. By late anaphase, *III*, the proteins of the actin-myosin cytoskeleton have assembled into the cleavage furrow and begin to exert a contractile force that divides cell, *IV*, and concurrently results in the disassembly of the cleavage furrow contractile apparatus. At the end of cytokinesis, *V*, two rings of actin and myosin and associated proteins constrict the distal ends of the midbody that connects the daughter cells, and stress fibers begin to reassemble in the cytoplasm. A muscle stem cell that differentiates, *VI*, may form myofibrils by the gradual exchange of muscle specific isoforms of the actin-myosin proteins with the nonmuscle isoforms in the stress fibers. Embryonic cardiomyocytes can undergo several rounds of cell division. Many of the same proteins that constitute the myofibrils are also used to form the contractile units in the cleavage furrows. In contrast to embryonic cardiomyocytes, skeletal myotubes are postmitotic cells. The three contractile systems, stress fibers, cleavage furrows, and myofibrils, are each attached to the cell membrane in complexes best understood for stress fibers and myofibrils, and yet to be understood for cleavage furrows. All three actin-containing structures (stress fibers, cleavage furrows, and myofibrils) may share not only a sarcomeric structure, but common means of attachment to their plasma membrane as well. (Reproduced with permission of Cell Motility and the Cytoskeleton)

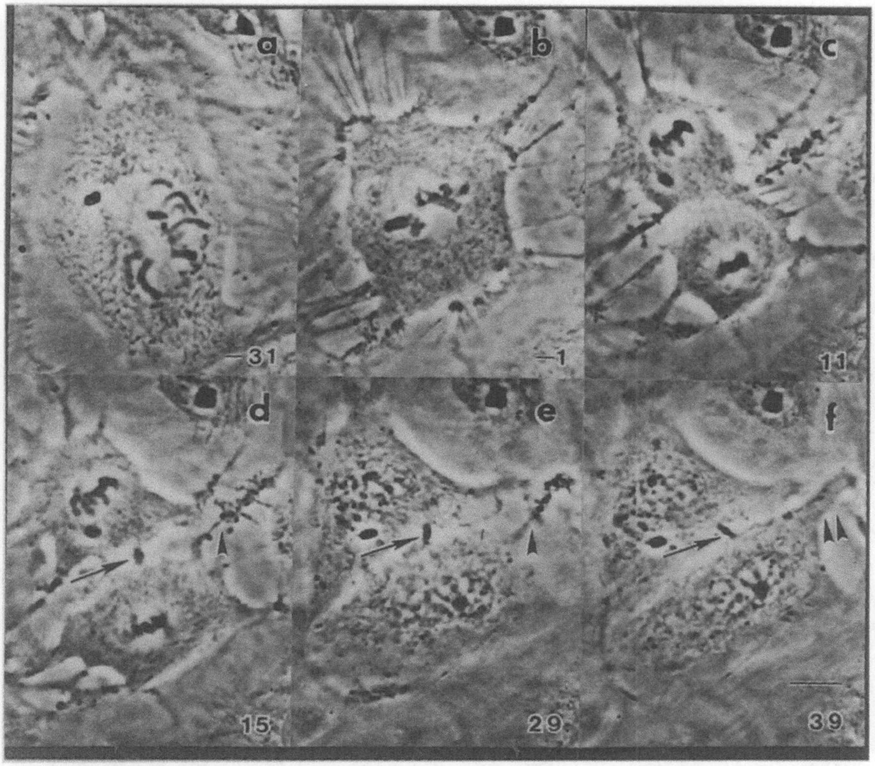

Fig. 2 a–f. Phase-contrast video micrographs of a dividing PtK2 cell that summarize the contractile activity and changes in cell shape that can be correlated with the changing patterns of actin and myosin seen during division in microinjected cells. During prophase, **a**, the cell begins to shorten and round up. At the same time, stress fibers shorten and begin to disassemble. At the time of metaphase, **b**, the cell is shorter and thicker and contains a mitotic spindle that excludes cytoplasmic organelles. Stress fiber disassembly provides a pool of actin and myosin that is concentrated in the spindle. During cytokinesis, **c**, additional cell rounding occurs together with a contraction of the cleavage furrows where actin and myosin are concentrated. At the end of cytokinesis, **d–f**, a tight mid-body (*arrow*) forms and undergoes rocking movements and changes in orientation (compare **e** with **f**). Gaskets of actin, filamin and myosin are present at either end of the mid-body, which becomes more constricted while retraction fibers (*single arrowheads* in **d** and **e**) that formed when the cell rounded up. refill with cytoplasm, and spread out (*double arrowheads* in **f**). Actin and myosin reassemble in stress fibers that are first seen in association with the mid-body gaskets. *Numbers* are minutes before (–) and after the onset of anaphase. *Bar* 10 μ. (Reproduced with permission of Cell Motility and the Cytoskeleton)

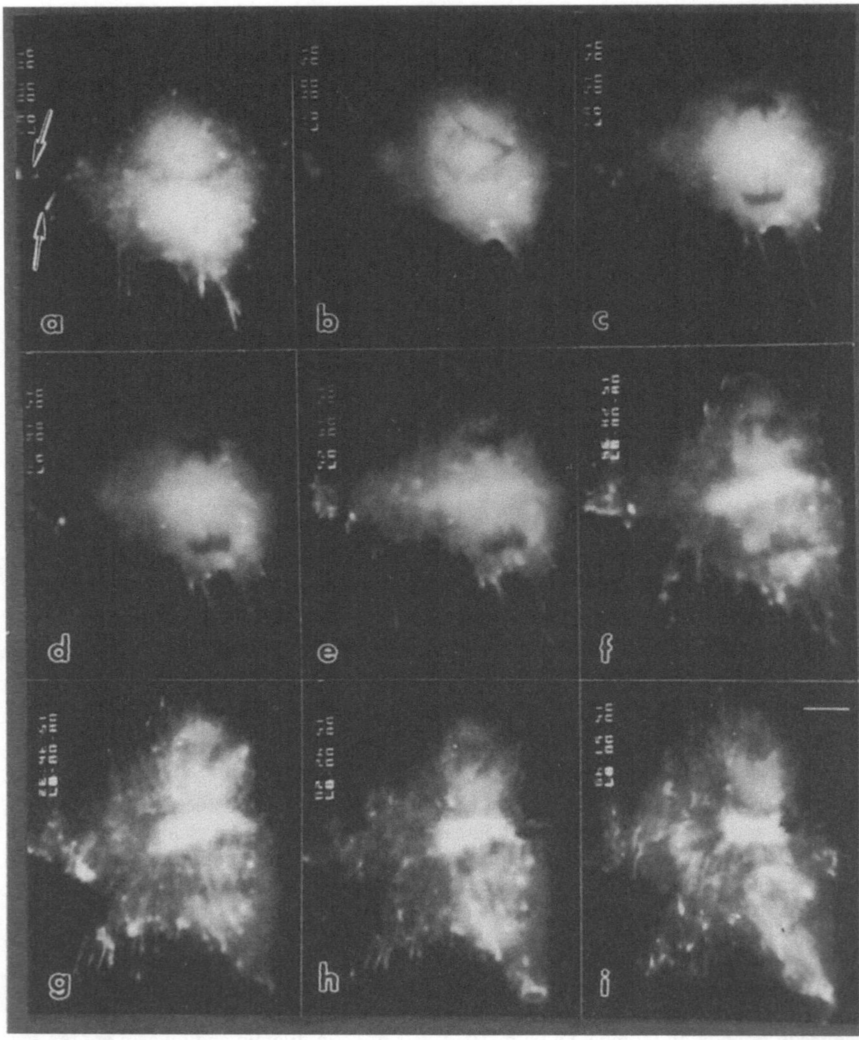

Fig. 3 a–i. A PtK2 cell that has been injected previously with fluorescently labeled actin. At metaphase, **a**, 20 min before the onset of anaphase, the cell shape is eccentric with an extended left lobe. The *arrows* indicate the margin of the out-of-focus lobe. In this higher focal plane, actin is concentrated in the mitotic spindle. **b** and **c** One and 7 min, respectively, after the onset of anaphase, actin remains concentrated in the interzonal region of the spindle. One minute later, **d**, the first indication of an equatorial band of actin appears on the surface beneath the former metaphase plate. **e** and **f** Three and 9 min later, respectively, the actin equatorial band becomes brighter and wider. However, the actin band does not extend across the left lobe of the cell causing an asymmetrical contraction (**g** to **i**). (Time after the onset of anaphase; **e** 11 min, **f** 20 min, **g** 28 min, **h** 34 min, **i** 43 min). Reforming stress fibers radiate out from either side of this contracting equatorial as cell spreading is occurring prematurely in this cell. Eventually, the furrow receded with a concomitant disassembly of the band of actin, resulting in the formation of a binucleate cell. *Bar* 10 μ. (Reproduced with permission of Cell Motility and the Cytoskeleton)

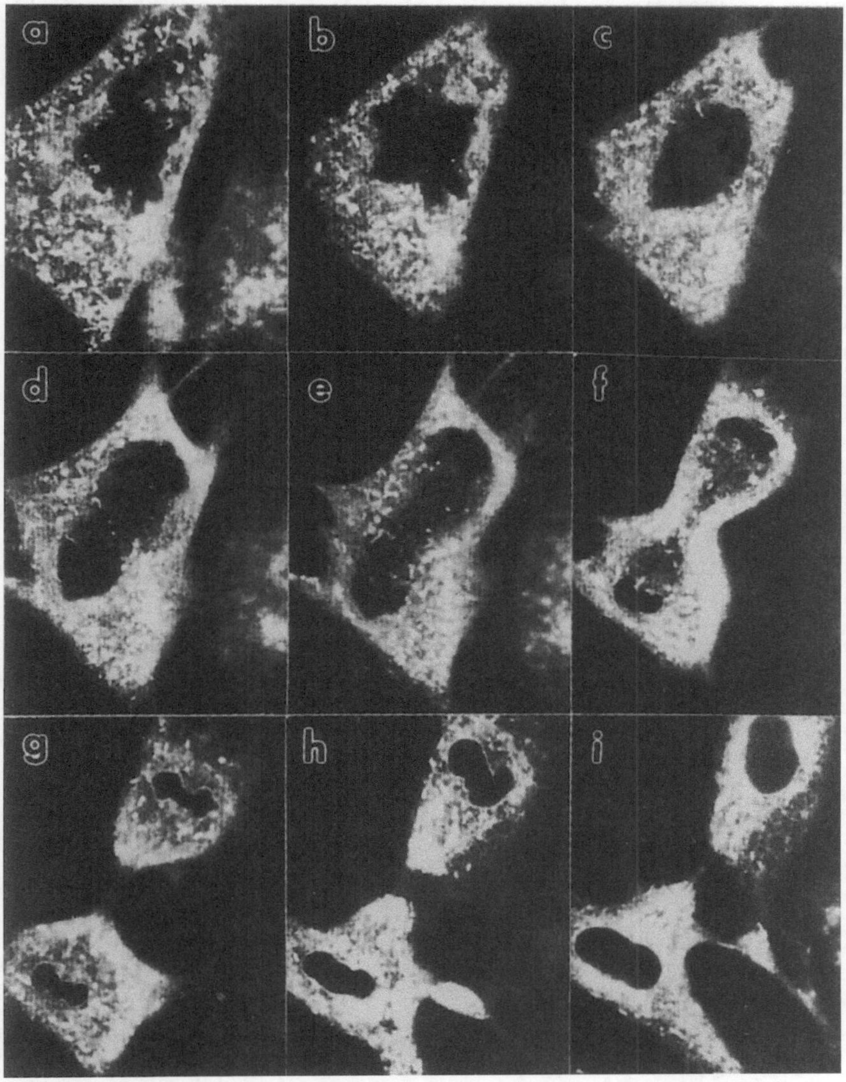

Fig. 4 a–i. A series of time points, beginning at metaphase, of optical sections through the central plane of a DiOC6(3)-stained PtK2 cell. Note the absence of membrane staining in the metaphase spindle in **a** and **b**. As the chromosomes separate (**c** to **e**), they leave in their wake a membrane void region. With the onset of anaphase B, membranous figures begin to encroach on the interzone (**e**), and are compressed and concentrated there as the contractile ring begins cytokinesis (**f** to **h**). As the daughter cells spread, the lace-like endoplasmic reticulum network is reestablished in the thin regions of the newly formed cells. Elapsed time: **a** 0 min 00 s, **b** 9 min 3 s, **c** 12 min 38 s, **d** 15 min 31 s, **e** 18 min 20 s, **f** 21 min 13 s, **g** 23 min 49 s, **h** 29 min 40 s, **i** 40 min 49 s. *Bar* 5 µ. (Reproduced with permission of Cell Motility and the Cytoskeleton)

1986; Rhee et al. 1994). Fluorescence microscopy of living muscle cells injected with fluorescent labeled alpha-actinin allowed us to follow the growth of the short linear spacings of the premyofibril (Rhee et al. 1994) to the long spacings of the alpha-actinin containing Z-bands of the mature myofibrils (Sanger et al. 1986). Similar microscopic methods have shown how the infectious bacterium, *Listeria monocytogenes*, moves inside infected PtK2 cells (Sanger et al. 1992; Dold et al. 1994; Nanavati et al. 1994; Sanger et al., in Press). Actin and actin binding proteins are conscripted by the infectious bacterium and used to support locomotion of the bacterium in the cytoplasm of the infected cell (Fig. 5). It was experiments on living cells that determined that the *Listeria* moved away from the newly formed tail and that new monomer actin molecules were incorporated into the tails at the bacterial surfaces (Dabiri et al. 1990; Sanger et al. 1992; Theriot et al. 1992).

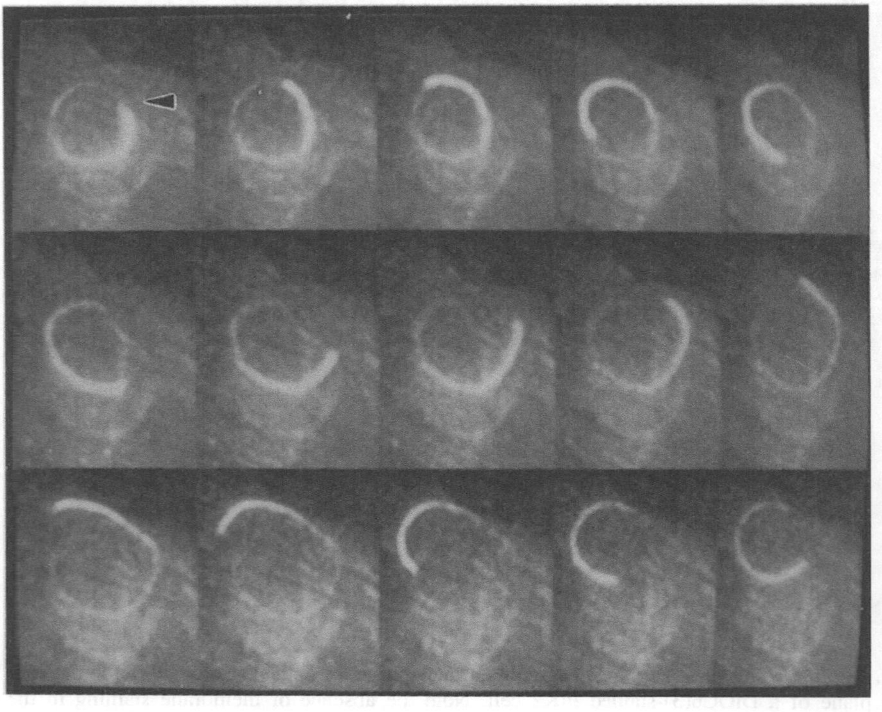

Fig. 5. This montage illustrates the pathway taken by the infectious pathogen, *Listeria monocytogenes*, in the cytoplasm of an infected PtK2 cell that had been microinjected with fluorescently labeled alpha-actinin. The position of the bacterium was determined by phase-contrast microscopy and is located just in front of the leading edge of the fluorescent alpha-actinin containing tail (*arrowhead*). *Each image* represents a time point 25 s after the previous image, *moving from left to right beginning with the top row*. (Reproduced with permission of Cell Motility and the Cytoskeleton)

2 Preservation and Availability of Experimental Films and Videotapes

Studies of the motility of cells, organelles, and their component molecules clearly require a variety of techniques to dissect the underlying mechanisms that produce the movement. One key to understanding and appreciating cell motility is the methodology devised for visualizing and recording the motile events as they occur in whole cells or cell-free fractions. Early studies used film to record events that could be detected by microscopic modes such as phase-contrast or polarization optics. The advent of video microscopy and digital image processing provided an increased capacity to detect the dynamic events of cell motility, and also provided a more versatile medium, video tape and video disk, for recording and displaying data. Papers documenting the observations are recorded in a printed medium in various journals. A few photographic images are presented along with the printed sentences to convey the motility that was analyzed. What is missing in the printed medium and static photographs are not only the multiple motile events that can only be appreciated in the moving images, but also the inherent elegance and beauty of the motile events that are so characteristic of living systems. How can these moving images be made available to other investigators as well as to the future investigators, i.e., the students?

Through the efforts of the late Robert D. Allen, editor of the journal, Cell Motility, a video disk supplement to that journal was published in 1983 so that a wide audience could view the recorded sequence of cell motility on which experiments were based. Increasing numbers of investigators now publish original observations in the successor to that videodisc, the Video Supplement of the journal, Cell Motility and the Cytoskeleton (Sanger and Sanger 1992a, b; 1994a, b) which is published on videotape. In addition, one of the first video supplements was devoted to sequences of notable examples of microtubular-based motility that were recorded between 1950 and 1990 (Sanger and Sanger 1990a, b), and a similar volume on actin/myosin-based motility is in preparation. At this meeting, we presented selected highlights of the past and future video supplements to illustrate the variety of approaches that have been undertaken to study the properties of motility in cells and the molecular motors responsible for them.

References

Allen RB (1983) Cell motility. Videodisc supplement 1

Dabiri G, Sanger JM, Portnoy DA, Southwick FS (1990) Listeria monocytogenes moves rapidly through the host-cell cytoplasm by inducing directional actin assembly. Proc Natl Acad Sci USA 87:6068–6072

Dold F, Sanger JM, Sanger JW (1993) Intact alpha-actinin molecules are needed for both the assembly of actin into the tails and the locomotion of *Listeria monocytogenes* inside infected cells. Cell Motil Cytoskel 28:97–107

Nanavati D, Ashton FT, Sanger JM, Sanger JW (1994) The dynamics of actin and alpha-actinin in the tails of *Listeria monocytogenes* in infected cells. Cell Motil Cytoskel 28:346–358

Rhee D, Sanger JM, Sanger JW (1994) The premyofibril: evidence for its role in myofibrillogenesis. Cell Motil Cytoskel 28:1–24

Sanger JM, Dome JS, Hock RS, Mittal B, Sanger JW (1994) The occurrence of contractile fibers and their association with talin in the cleavage furrow in living PtK2 cells. Cell Motil Cytoskel 27:26–40

Sanger JM, Dome JS, Mittal B, Somlyo AV, Sanger JW (1989b) Dynamics of the endoplasmic reticulum in living non-muscle and muscle cells. Cell Motil Cytoskel 13:301–319

Sanger JM, Mittal B, Dome JS, Sanger JW (1989a) Analysis of cell division using fluorescently labeled actin and myosin in living PtK2 cells. Cell Motil Cytoskel 14:201–219

Sanger JM, Mittal B, Southwick FS, Sanger JW (1995) Listeria monocytogenes intracellular migration: inhibition by profilin, vitamin D binding protein and DNase I. Cell Motil Cytoskel 30:38–49

Sanger JM, Sanger JW (1990a) Cell motility and the cytoskeleton. Video Supplement 2

Sanger JM, Sanger JW (1990b) Video Supplement 2. Video tape contents, narration and references. Cell Motil Cytoskel 17:356–372

Sanger JM, Sanger JW (1992a) Cell motility and the cytoskeleton. Video Supplement 3

Sanger JM, Sanger JW (1992b) Video Supplement 3. Video tape contents, narration and references. Cell Motil Cytoskel 23:71–82

Sanger JM, Sanger JW (1994a) Cell motility and the cytoskeleton. Video supplement 4

Sanger JM, Sanger JW (1994b) Video supplement 4. Video tape contents, narration and references. Cell Motil Cytoskel 27:361–372

Terasaki M (1989) Fluorescent labeling of endoplasmic reticulum. Methods Cell Biol 29:125–135

Theriot JA, Mitchison TJ, Tilney LG, Portnoy DA (1992) The rate of actin-based motility of intracellular *Listeria monocytogenes* equals the rate of actin polymerization. Nature (London) 357:257–260

Waterman-Storer CM, Sanger JM, Sanger JW (1993) Dynamics of organelles in the mitotic spindles of living cells: membranes and microtubule interactions. Cell Motil Cytoskel 26:19–39

Actin Polymerization by the Intracellular Bacterial Pathogen, *Listeria monocytogenes*

P. COSSART[1]

1 Introduction

L. monocytogenes is a bacterial pathogen discovered relatively recently (Murray 1926). It is responsible for severe human food-borne infections leading to meningitis, meningo-encephalitis, septicemias, abortions, and death in 30% of the cases. It also naturally infects many other animal species, including cows and sheep, and is therefore of veterinary importance. Most of our knowledge of the human disease comes from the many studies carried out in mice.

Via contaminated food, bacteria reach the gastrointestinal tract and cross the intestinal barrier. They are subsequently engulfed by macrophages. Then, via the lymph and the blood, they reach the liver and the spleen, where most are killed by nonspecific host defense mechanisms. Depending on the immune response of the host, remaining bacteria will either be eliminated or multiply unrestrictedly in the hepatocytes, from which they will disseminate via the blood, to the brain and the placenta. It was shown in the 1960s by Mackaness that recovery from infection and protection against secondary infection require the induction of a specific T cell-dependent immune reponse, antibodies playing no role in the clearing of the bacteria (Mackaness 1962). Following this pioneering work, *L. monocytogenes* has become a model system to study the induction of the immune T-cell response (Kaufmann 1993), and research in this area has culminated recently with the discovery of the first bacterial protective CD8+ epitope (Harty and Bevan 1992).

L. monocytogenes, as also shown by Mackaness, can survive and replicate in macrophages but can also induce its own phagocytosis by normally nonphagocytic cells. It has become a model system to study intracellular parasitism by invasive bacteria (Cossart and Mengaud 1989). This property is mainly due to three factors: (1) the possibility to infect cells in vitro, (2) the availability of genetic tools, (3) the existence of a "natural animal model", the murine infection. All three factors have contributed to a recent accumulation of data concerning the strategies and the molecular determinants used by the bacterium to infect cells and tissues (reviewed in Portnoy et al. 1992; Tilney and Tilney 1993; Cossart 1994; Cossart and Kocks 1994; Sheehan et al. 1994).

---

[1] Unité des Interactions Bactéries-Cellules, Institut Pasteur, 28 rue du Dr. Roux, Paris 75015, France.

45. Colloquium Mosbach 1994
The Cytoskeleton

2 For L. monocytogenes, Intracellular Actin-Based Motility and Direct Cell-to-Cell Spread Are Key Virulence Determinants

The cell infectious process can be artificially divided into four steps: entry into a cell, escape from the phagocytic vacuole, intracytoplasmic multiplication, and cell-to-cell spread.

Very soon after the interaction with a cell, the listeriae are internalized and reside within membrane-bound vacuoles during a very short period, i.e., 30 min. Then, the bacteria lyse the vacuoles and reach the cytosol. In this environment, bacteria start to multiply (at about the same rate as in broth medium, i.e., a doubling time of 1 h (Marquis et al. 1993). Concomittantly, the become covered with actin filaments (Tilney and Portnoy 1989; Mounier et al. 1990) which, within 2 h, rearrange in long comet tails left behind in the cytosol while bacteria are moving within the cytosol at a speed of about 0.3 $\mu\ s^{-1}$ (Fig. 1; Dabiri et al. 1990; Sanger et al. 1992; Theriot et al. 1992). Observation of thin cross sections of *Listeria*-infected cells decorated with fragment S1 of myosin reveals that the actin tails are made of cross-linked short filaments with their barbed (fast polymerizing) ends towards the bacterium, suggesting that actin polymerization takes place at the rear of the bacterium (Tilney and Portnoy 1989). Interestingly, the actin filaments in the tail are not aligned and appear somewhat disorganized. Microinjection of fluorescent actin monomers in live infected cells has indeed demonstrated that the actin polymerisation takes place at the rear of the bacterium (Sanger et al. 1992). Other studies have shown that the speed of movement is strictly corrolated to the rate of actin polymerization at the posterior end of the bacterium (Theriot et al. 1992). The direction of this intracellular movement appears random although the bacteria are mainly moving in pseudocircular movements

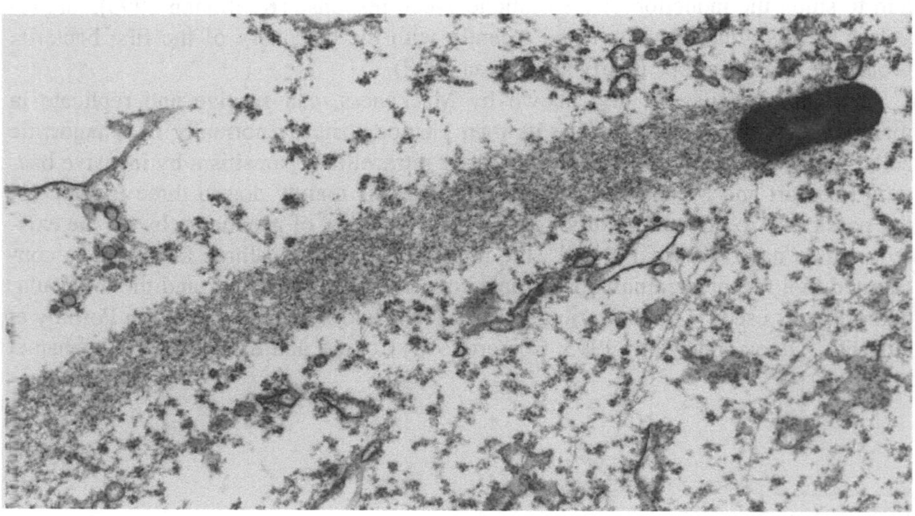

Fig. 1. Electron micrograph of a J774 cell infected with *L. monocytogenes* showing an actin comuet tail behind an intracellular bacterium. (Kocks et al. 1992)

(Theriot et al. 1992; Nanavati et al. 1994). It is to be noted that a single bacterium can move and stop within very short periods of time, and that the speed of movement varies between bacteria. It also varies with time for one given bacterium.

When the bacteria reach the plasma membrane, they induce the formation of long protrusions which contain one bacterium at their tips (Tilney and Portnoy 1989; Mounier et al. 1990). These protuberances invaginate in the neighboring cells and are subsequently phagocytosed, giving rise to two-membrane vacuoles which in turn are lysed, liberating the bacteria in the cytosol of a second infected cell. By this mechanism, *Listeria* can spread directly from cell to cell and disseminate within tissues. This strategy allows the bacteria to be protected from host defenses such as circulating antibodies. Mutants which are unable to move or spread are strongly affected in virulence (Domann et al. 1992; Kocks et al. 1992; Goossens and Milon 1992; Brundage et al. 1993).

3 The Actin-Based-Motility of *L. Monocytogenes*

3.1 Contribution of the Bacterium: the actA Gene Product

Through the study of a mutant unable to spread from cell to cell, (i.e., unable to form plaques on a cell monolayer), we have identified the first *Listeria* gene involved in the actin polymerization process (Kocks et al. 1992). This gene is called *actA*. It en-

```
-29 MGLNRFMRAMMVVFITANCITINPDIIFA
    ******
  1 ATDSEDSSLNTDEWEEEKTEEQPSEVNTGPRYETAREVSSRDIKE
 46 LEKSNKVRNTNKADLIAMLKEKAEKGPNINNNNSEQTENAAINEE
 91 ASGADRPAIQVERRHPGLPSDSAAEIKKRRKAIASSDSELESLTY
136 PDKPTKVNKKKVAKESVADASESDLDSSMQSADESSPQPLKANQQ
181 PFFPKVFKKIKDAGKWVRDKIDENPEVKKAIVDKSAGLIDQLLTK
226 KKSEEVNAS
235 DFPPPPTDEELRLALPETPMLLGFNAPATSEPSSF
270 EFPPPPTDEELRLALPETPMLLGFNAPATSEPSSF
305 EFPPPPTEDELEIIRETASSLDSSFTRGDLASLRNAINRHSQNFS
350 DFPPIPTEEELNGRGGR
367      PTSEEFSSLNSG
379 DF     TDDENSET
389        TEEEIDRLADLRDRGTGKHSRNAGFLPLNPFASSPVPSL
428 SPKVSKISAPALISDITKKTPFKNPSQPLNVFNKKTTTKTVTKKP
473 TPVKTAPKLAELPATKPQETVLRENKTPFIEKQAETNKQSINMPS
518 LPVIQKEATESDKEEMKPQTEEKMVEESESANNANGKNRSAGIEE
563 GKLIAKSAEDEKAKEEPGNHTTLILAMLAIGVFSLGAFIKIIQL
607 RKNN
```

Fig. 2. Amino-acid sequence of ActA. (Vasquez-Boland et al. 1992; Kocks et al. 1992)

codes a 610 amino-acid surface protein (Vasquez-Boland et al. 1992; Domann et al. 1992; Fig. 2). The two main features of this protein are the presence of proline-rich tandem repeats and a C-terminal region whose hydrophobicity suggests that it could be a membrane anchor. Homology searches in data banks has revealed that ActA is similar to caldesmon (20% similarity on a 243 overlap) and to microtubule associated protein 1(Map1) (20% similarity on a 196 overlap). Several peptides in the ActA sequence are reminiscent of peptides present in known actin binding proteins, like vinculin. However, the overall primary structure of the protein is quite unique.

The obtain insight into the mode of actin of actA, antibodies were raised against a synthetic peptide corresponding to the first proline-rich repeat, and infected cells were examined by double immunofluorescence with both affinity-purified anti-ActA antibodies and FITC-phallodin to label the polymerized actin (Kocks et al. 1993). ActA is not released in the comet tail, as also shown by Niebuhr et al. (1993). It is localized on the surface of the bacteria. However, ActA is unevenly distributed on the surface of the bacteria, with maximal amounts at the end located close to the comet tail. ActA is absent at one extremity of the bacterial body. The distribution of ActA on the bacterial surface strictly colocalizes with the distribution of actin, even before comet tail formation. These results indicate that the distribution of ActA predetermines the site of actin assembly. Moreover, the absence of ActA at one pole predetermines the direction of movement in the direction of the non-ActA expressing pole.

In order to assess the role of ActA in the process, the *actA* gene was expressed "outside" from the genetic context of *L. monocytogenes*. Two types of heterologous expression of ActA were performed.

1. ActA was expressed in *Listeria innocua*, a bacterium which exclusively lives in the environment (Kocks et al., in prep.). To test the effect of the ActA expression, the recombinant *Listeria* were incubated in *Xenopus Laevis* oocytes extracts which represent a very convenient system to test bacterial actin-based motility, *in vitro* (Theriot et al. 1994). In these extracts, provided fluorescent monomeric actin is added, actin tails can be vizualized just as if the bacteria were in the cytosol of an infected cell. We have shown that *L. innocua* expressing ActA is able to induce actin polymerization and comet tail formation.
2. The *actA* gene was transfected and expressed in eukaryotic cells. Two types of constructs were obtained. (Friedrich et al., in prep.). One of them was devoid of the N-terminal signal sequence and of the C-terminal membrane anchor encoding regions, resulting in a soluble form of ActA. The other construct was fused to a CAAX motif-encoding fragment in order to target ActA to the plasma membrane. Production of a soluble ActA protein clearly induces a strong actin polymerization as vizualized by immunofluorescence. Expression of the membrane form of ActA leads to preferential actin polymerization at the plasma membrane and induces aberrant membrane deformations. These results are in agreement with those obtained by Pistor et al., who showed that ActA, when expressed with its membrane anchor, induces actin polymerization within eukaryotic cells, and interestingly targets to mitochondria (Pistor et al. 1994).

3.2 Contribution of Other Cellular Factors

Immunofluorescence techniques have demonstrated that two actin cross-linking proteins, alpha-actinin (Dabiri et al. 1990; Sanger et al. 1992) and fimbrin/plastin (Kocks and Cossart 1994), colocalize with the tail. Tropomyosin, an actin-binding protein which stabilizes the actin filaments also colocalizes with the actin tail (Dabiri et al. 1990). Evidence for a functional role of alpha-actinin has recently been provided (Dold et al. 1994). It is not yet the case for fimbrin or tropomyosin.

Recent exciting findings have been published by Theriot et al. (1994), who demonstrated that profilin colocalizes with the site of actin assembly. Some of the immunofluorescence pictures of infected cells labeled with antiprofilin antibodies indicate that the labeling of profilin is quite similar to that of ActA. In addition, in *Xenopus laevis* oocytes extracts that have been depleted in profilin, bacterial movement is strongly impaired (Theriot et al. 1994). However, to date, no direct interaction between ActA and profilin has been presented.

4 Putative Models

Bacteria grown in broth are unable to polymerize actin in vitro (Tilney et al. 1992). This observation led to the hypothesis that ActA should be modified inside cells to interact with the cytoskeleton. This modification was anticipated to be a true covalent modification or the recruitment of a nucleator, or both. It has been recently demonstrated that ActA is phosphorylated in vivo (Brundage et al. 1993) and can be phosphorylated in vitro by casein kinase II (Kocks et al. 1993), but whether this phosphorylation is sufficient to induce the actin assembly process is unknown. Therefore, whether ActA is or recruits a nucleator remains an open question, as shown in Fig. 3. In both cases, ActA could, through its proline-rich repeats, attract the profilin-actin complexes which would then be added to the nascent free barbed ends. It is also possible that profilin colocalizes only with the region of actin assembly due to the presence of free barbed ends which have high affinity for the profilin-actin complexes (Pantaloni and Carlier 1994; Pring et al. 1992). After nucleation and initial polymerisation, filaments are released and cross-linked. Attachment (cross-linking) of these filaments to other new filaments or to the performed meshwork and concomittant new nucleation events on the bacterial surface would propel the bacterium in the direction of the non-ActA expressing pole.

5 Concluding Remarks

A few other unrelated pathogens use a similar mechanism of continuous actin assembly to move intracellularly (Karunasagar et al. 1993; Bernardini et al. 1989; Heinzen et al. 1993); Teyssière et al. 1992). Interestingly, the protein identified in *Shigella flexneri*, which seems to play the same role as ActA, IcsA, does not share any simi-

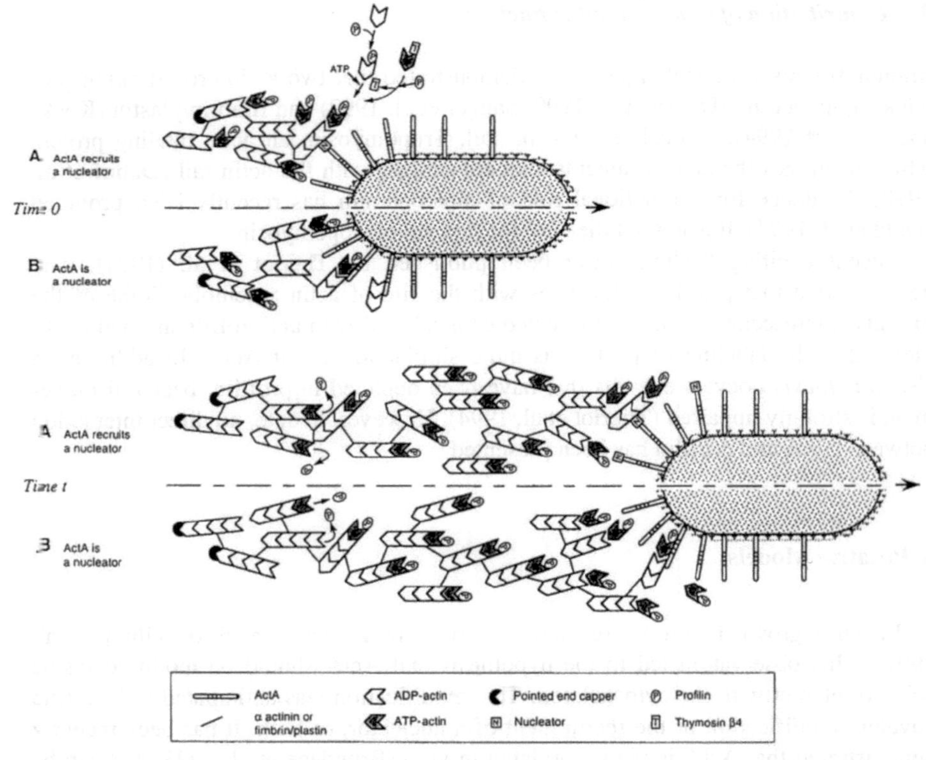

Fig. 3. Putative model for actin polymerization by *L. monocytogenes*. (Cossart and Kocks 1994)

larity with ActA at the amino-acid sequence level, and does not contain the characteristic proline-rich regions (Bernardini et al. 1989; Goldberg et al. 1993; Lett et al. 1989). These observations suggest that the next important steps in our understanding of the bacterial actin-based motility will come from biochemical or other studies leading to the identification of the factors interacting with ActA or IcsA.

References

Bernardini ML, Mounier J, D'Hauteville H, Coquis-Rondon M & Sansonetti PJ (1989). Identification of *icsA*, a plasmid locus of *Shigella flexneri* that governs bacterial intra- and intercellular spread through interaction with F-actin. Proc Natl Acid Sci USA 86:3867–3871

Brundage RA, Smith GA, Camilli A, Theriot JA & Portnoy DA (1994) Expression and phosphorylation of the *Listeria monocytogenes* ActA protein in mammalian cells. Proc Natl Acad Sci USA 90:11890–11894

Cossart P (1994). *Listeria monocytogenes*: strategies for entry and survival in cells and tissues. In: Clinical Baillère's Infectious Disease

Cossart P & Kocks C (1994). The actin based motility of the intracellular pathogen *Listeria monocytogenes*. Mol Microbiol in press

Cossart P & Mengaud J (1989). *Listeria monocytogenes*: a model system for the molecular study of intracellular parasitism. Mol Biol Med 6:464–474

Dabiri GA, Sanger JM, Portnoy DA and Southwick FS (1990) *Listeria monocytogenes* moves rapidly through the host-cell cytoplasm by inducing directional actin assembly. Proc Natl Acad Sci USA 87:6058–6072

Dold FG, Sanger JM & Sanger JW (1994) Intact alpha-actinin molecules are needed for both the assembly of actin into tails and the locomotion of *Listeria monocytogenes* inside infected cells. Cell Mot Cytoskel 28:97–107

Domann E, Wehland J, Rohde M, Pistor S, Hartl M, Goebel W, Leimeister-Wächter M, Wuenscher M & Chakraborty T (1992) A novel bacterial gene in *Listeria monocytogenes* required for host cell microfilament interaction with homology to the proline-rich region of vinculin. EMBO J 11:1981–1990

Goldberg M, Bârzu O, Parsot P & Sansonetti PJ (1993) Unipolar localization and ATPase activity of IcsA, a *Shigella flexneri* protein involved in intracellular movement. Bacteriol 175:2189–2196

Goossens P & Milon G (1992) Induction of protective CD8+ T lymphocytes by an attenuated *Listeria monocytogenes actA* mutant. Int-Immunol 4:1413–8

Harty JT & Bevan M (1992) CD8+ T cells specific for a single nonamer epitope of *Listeria monocytogenes* are protective in vivo. J Exp Med 175:1531–1538

Heinzen RA, Hayes SF, Peacock MG & Hackstadt T (1993) Directional actin polymerization associated with spotted group *Rickettsia* infection of vero cells. Infect Immun 61:1926–1935

Karunasagar I, Krohne G & Goebel W (1993) *Listeria ivanovii* is capable of cell-to-cell spread involving actin polymerization. Infect Immun 61:162–169

Kaufmann SHE (1993) Immunity to intracellular bacteria. Ann Rev Immunol 11:129–163

Kocks C, Gouin E, Tabouret M, Berche P, Ohayon H & Cossart P (1992) *Listeria monocytogenes*-induced actin assembly requires the *actA* gene product, a surface protein. Cell 68:521–531

Kocks C, Hellio R, Gounon P, Ohayon H & Cossart P (1993) Polarized distribution of *Listeria monocytogenes* surface protein ActA at the site of directional actin assembly. J Cell Sci 105:699–710

Lett MC, Sasakawa C, Okada N, Sakai T, Makino S, Yamada M, Komatsu K & Yoshikawa M (1989) *virG*, a plasmid-coded virulence gene of *Shigella flexneri*: identification of the virG protein and determination of the complete coding sequence. J Bacteriol 171:353–359

Mackaness GB (1962) Cellular resistance to infection. J Exp Med 116:381–406

Marquis H, Bouwer HA, Hinrichs D & D, Portnoy DA (1993) Intracytoplasmic growth and virulence of *Listeria monocytogenes* auxotrophic mutants. Infect Immun 61:3756–2760

Mounier J, Ryter A, Coquis-Rondon M & Sansonetti PJ (1990) Intracellular and cell-to-cell spread of *Listeria monocytogenes* involves interaction with F-actin in the enterocyte-like cell line Caco-2. Infect Immun 58:1048–1058

Murray EGD, Webb RE & Swann MBR (1926) A disease of rabbits characterized by a large mononuclear leucocytosis, caused by a hitherto undescribed bacillus *Bacterium monocytogenes* (n. sp.). J Pathol Bacteriol 29:407–439

Nanavati DFT, Ashton FT, Sanger J & Sanger J (1994) Dynamics of actin and alpha-actinin in the tails of *Listeria*-infected PtK2 cells. Cell Mot Cytoskelet 28: ..

Niebuhr K, Chakraborty T, Rohde M, Gazlig T, Jansen B, Kollner P & Wehland J (1993) Localization of the ActA polypeptide of *Listeria monocytogenes* in infected tissue culture cell lines: ActA is not associated with actin comets. Infect Immun 61:2793–2802

Pantaloni D & Carlier MF (1993) How profilin promotes actin filament assembly in the presence of thymosin b4. Cell 75:1007–1014

Pistor S, Chakraborty T, Niebuhr K, Domann E & Wehland J (1994) The ActA protein of *L. monocytogenes* acts as a nucleator inducing reorganization of the actin cytoskeleton. EMBO J 13:758–763

Portnoy DA, Chakraborty T, Goebel W & Cossart P (1992 Molecular determinants of *Listeria monocytogenes* pathogenesis. Infect Immun 60:1263–1267

Pring M, Weber A & Bubb MR (1992) Profilin-actin complexes directly elongate actin filaments at the barbed end. Biochem 55:1827–1836

Sanger JM, Sanger JW & Southwick FS (1992) Host cell actin assembly is necessary and likely to provide the propulsive force for intracellular movement of *Listeria monocytogenes*. Infect Immun 60:3609–3619

Southwick FS & Purich DL (1994) Arrest of *Listeria* movement in host cells by a bacterial ActA analogue: implications for actin-based motility. Proc Natl Acad Sci USA 91:5168–5172

Teysseire NC, Chiche-Portiche C & Raoult D (1992) Intracellular movements of *Rickettsia conorii* and *R. typhi* based on actin polymerization. Res Microbiol 143:821–829

Theriot JA, Rosenblatt J, Portnoy DA, Goldschmidt-Clermont PJ & Mitchison T (1994) Involvement of profilin in the actin-based motility of *L. monocytogenes* in cells and in cell-free extracts. Cell 76:505–517

Tilney LG, Connelly PS & Portnoy DA (1990) Actin filament nucleation by the bacterial pathogen, *Listeria monocytogenes*. J Cell Biol 111:2979–2988

Theriot JA, Mitchison TJ, Tilney LG & Portnoy DA (1992) The rate of actin-based motility of *Listeria monocytogenes equals* the rate of actin polymerization. Nature 357:257–260

Tilney LG, DeRosier DJ & Tilney MS (1992) How *Listeria* exploits host cell actin to form its own cytoskeleton. I. Formation of a tail and how that tail might be involved in movement. J Cell Biol 118:71–81

Tilney LG, DeRosier DJ, Weber A & Tilney MS (1992) How *Listeria* exploits host cell actin to form its own cytoskeleton. II. Nucleation, actin filament polarity, filament assembly, and evidence for a pointed end capper. J Cell Biol 118:83–93

Tilney LG and Tilney MS (1993) The wily ways of a parasite: induction of actin assembly by *Listeria*. Trends in Microbiol 1:25–31

Tilney LG & Portnoy DA (1989) Actin filaments and the growth, movement, and spread of the intracellular bacterial parasite, *Listeria monocytogenes*. J Cell Biol 109:1597–1608

Vasquez-Boland JA, Kocks C, Dramsi S, Ohayon H, Geoffroy C, Mengaud J & Cossart P (1992) Nucleotide sequence of the lecithinase operon of *Listeria monocytogenes* and possible role of lecithinase in cell-to-cell spread. Infect Immun 60:219–230

Microtubules, Tau Protein, and Paired Helical Filaments in Alzheimer's Disease

E. MANDELKOW, J. BIERNAT, B. LICHTENBERG-KRAAG, G. DREWES, H. WILLE, N. GUSTKE, K. BAUMANN, and E.-M. MANDELKOW[1]

1 Introduction

The pathology of Alzheimer's disease (AD) can be recognized by two types of protein deposits, the amyloid plaques and the neurofibrillary deposits (tangles, neuropil threads). The latter are composed largely of paired helical filaments which are in turn made up mainly of an insoluble form of the microtubule-associated protein tau (Brion et al. 1985). The neurofibrillary deposits are particularly useful in defining the stages of AD progression because they proceed with a well-defined spatial and temporal pattern, starting with the transentorhinal region (stage 1) until eventually the frontal and temporal cortex are affected (stage 6; Braak and Braak 1991; Braak et al. 1994). Only the last three stages are clinically recognizable as AD, while the first three are not noticeable as such. This illustrates the problem in developing a cure or preventing the beginnings of the disease (for review see Braak and Braak 1994).

Microtubules are usually stabilized by cell-specific associated proteins (MAPs). In the case of neurons, the protein MAP2 is prominent in dendrites, while tau protein dominates in axons (Cleveland et al. 1977; Binder et al. 1985). The MAPs help to maintain axonal transport, hence they can be likened to "ties" that keep the microtubule "tracks" intact. Overexpression of tau leads to the formation of microtubule bundles and neurite extension (Kanai et al. 1992; Knops et al. 1991; Lo et al. 1993). Tau is a mixture of up to six isoforms in the human brain that arise from alternative splicing, and contain between 352 and 441 amino acid residues (Lee et al. 1988; Goedert et al. 1988, 1989; Himmler et al. 1989); in addition peripheral nervous tissue contains a "big tau" isoform (Couchie et al. 1992). Tau can be subdivided into several domains (Fig. 1), acidic, basic, proline-rich, repeats, pseudorepeat, and tail. The repeats are somehow involved in microtubule binding (Ennulat et al. 1989; Butner and Kirschner 1991), although the binding depends strongly on the flanking regions as well (Kanai et al. 1992; Brandt and Lee 1993; Gustke et al. 1994).

There are several modifications of tau in paired helical filaments: it is phosphorylated, aggregated, ubiquitinated, proteolytically processed, and it no longer binds to microtubules (for reviews see Anderton 1993; Goedert 1993; Mandelkow and Mandelkow 1993). Phosphorylation, aggregation, and detachment from microtubules appear to be early events in the abnormal transformation of tau, and are probably linked in some fashion. Ubiquitination (Bancher et al. 1991; Morishima-Kawashima et al.

[1] Max-Planck-Unit for Structural Molecular Biology, c/o DESY, Notkestrasse 85, D-22603 Hamburg, FRG. E-Mail MAND@MPASMB.DESY.DE.

45. Colloquium Mosbach 1994
The Cytoskeleton

Domain Structure of Tau

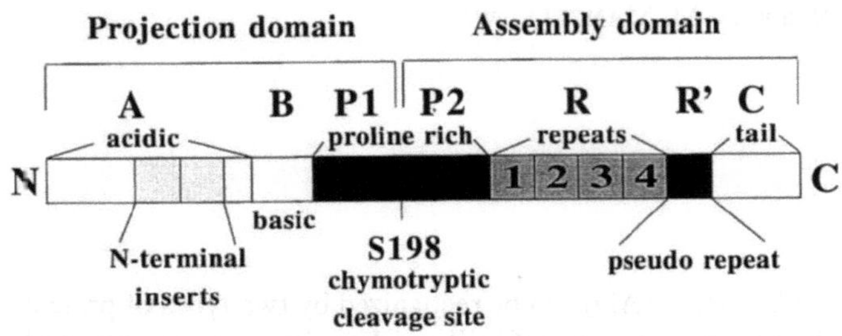

Fig. 1. Diagram of tau isoform htau40 (441 residues, Goedert et al. 1989). The two 29-residue inserts near the N-terminus are *lightly shaded*, the four repeats in the C-terminal half are *numbered 1–4 (medium shade)*. Definition of domains (Gustke et al. 1994): *projection domain* does not bind to microtubules by itself; *assembly domain* binds to microtubules, separable by chymotryptic cleavage. N-terminal domains *A* (acidic), *B* (basic), *P* (proline-rich and basic), separated into *P1* and *P2* at Y197. Repeats *R1–R2*, "5th" repeat *R'*; *C* C-terminal tail

1993) and proteolysis (Kondo et al. 1988; Novak et al. 1993) are probably secondary and may reflect the cell's attempt to get rid of the precipitated protein. Here we describe recent experiments in which we have addressed the following questions. Can we identify the differences between normal and pathological tau? Can we generate these differences in vitro and thereby study the causes of the pathological transformation? What are the structural and biochemical consequences of the transformation? What enzymes are involved? These questions were approached using a combination of biochemical, structural, and molecular biological methods which are described in the references in more detail.

2 Results and Discussion

2.1 *Detection of the Abnormal Phosphorylation of Tau*

In PHFs the phosphorylation of tau causes a reduced electrophoretic mobility (Grundke-Iqbal et al. 1986) so that one has to search for a protein kinase that would affect this parameter. Initially, we tested several well-known kinases. For example, CaM kinase induced a clear shift by phosphorylating a single site (Ser 416 in the htau40 numbering, see Steiner et al. 1990). A small shift can also be induced by PKA which phosphorylates Ser409 and several other sites (Scott et al. 1993, Biernat et al. 1993). However, the phosphorylated protein did not react with the PHF-specific antibodies. Other kinases such as PKC or cdc2 were even less effective with regard to electrophoretic mobility of PHF-antibodies (Steiner et al. 1990; Drewes et al. 1992).

On the other hand, one can argue that the kinase(s) that transform tau to an abnormal state are likely to be present in normal brain tissue. We therefore prepared brain

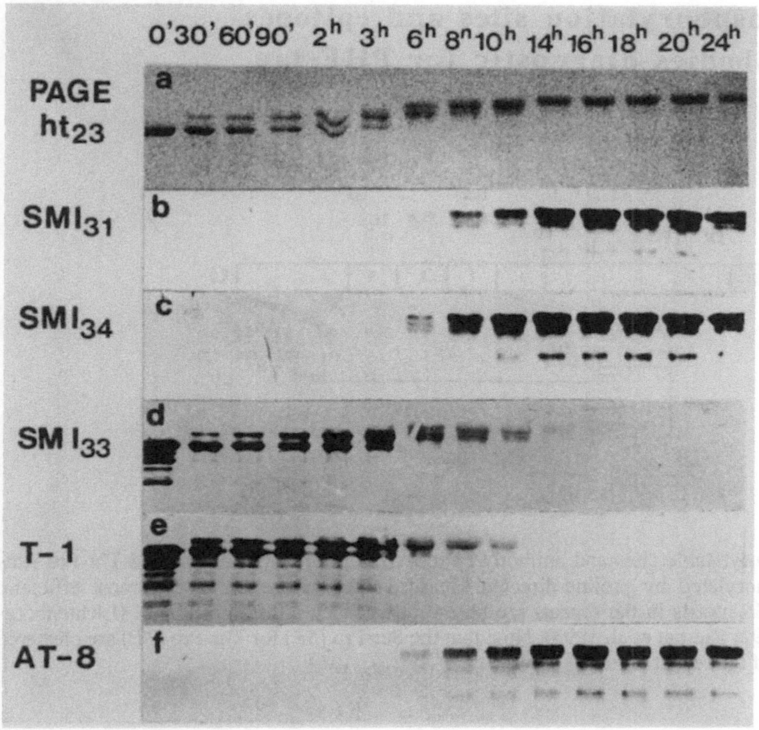

Fig. 2 a–f. Time course of phosphorylation of htau23 by kinase activity prepared from brain extract. **a** SDS-PAGE. **b–e** Immunoblots with several antibodies (Lichtenberg-Kraag et al. 1992). There is a progressive shift in M_r and a change in reaction with antibodies that distinguish between normal and abnormal tau. Antibodies SMI31, SMI34, and At-8 react with phosphorylated epitopes containing Ser-Pro motifs, antibodies SMI33, and Tau-1 react with unphosphorylated epitopes. Note that the kinase activity persists for up to 24 h

extracts and tested them for their ability to phosphorylate tau, with limited success. However, the failure to obtain phosphorylation could be due either to underphosphorylation or to overdephosphorylation. After all, normal brain tau is not pathologically phosphorylated, so that if there are "normal" kinases, then these must be balanced by "normal" phosphatases. Thus, abnormal phosphorylation could be expected only when the phosphatases were inhibited.

Phosphatases PP-1 and PP-2a are inhibited by the inhibitor okadaic acid (Bialojan and Takai 1988). Using this, we obtained a kinase activity from bran extract which was capable of conferring Alzheimer-like characteristics to tau protein. The kinase activity incorporated up to six phosphates into tau, caused a mobility shift in brain tau as well as all recombinant tau isoforms, and induced an Alzheimer-like antibody reactivity with several antibodies (Biernat et al. 1992; Lichtenberg-Kraag et al. 1992). All antibodies that discriminated between PHF tau and normal tau were phosphorylation sensitive, either in a positive sense (i.e., reacting with phosphorylated epitopes of PHFs and of recombinant tau, for example AT8 and SMI31), or in a negative sense

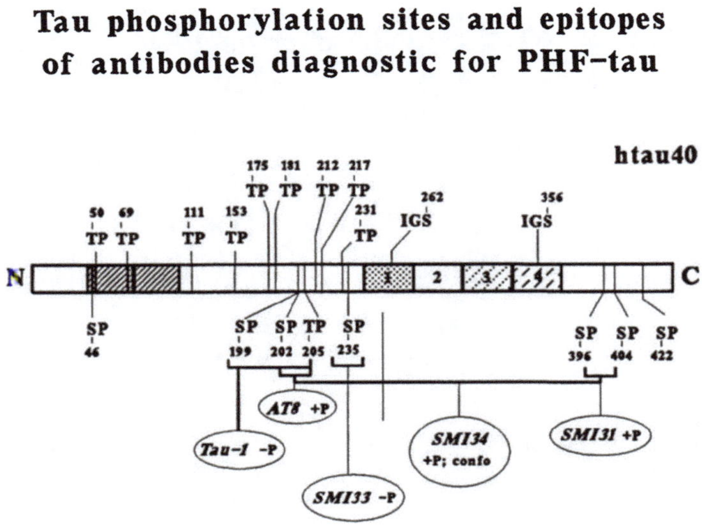

Fig. 3. Phosphorylatable sites and antibody epitopes in tau protein. Ser-Pro and Thr-Pro sites can be phosphorylated by proline-directed kinases, MAP kinase being that most efficient, whereas the IGS motifs in the repeats are phosphorylated by a different kinase (Lichtenberg-Kraag et al. 1992; Biernat et al. 1993). Note that the Ser-Pro *(SP)* for Thr-Pro *(TP)* are clustered on both sides of the repeat domain

(i.e, reacting only with normal and unphosphorylated tau, for example TAU1 and SMI33, see Fig. 2). These results proved that the Alzheimer-like kinase activity was already present in normal brain and not generated by a pathological condition.

In order to determine the epitopes of the antibodies and the phosphorylation sites which influence their reactivities, we used a combination of phosphopeptide sequencing (Meyer et al. 1991) and site directed mutagenesis. Most epitopes involved phosphorylatable serines followed by prolines (Fig. 3). An example is the epitope of the PHF-specific antibody AT8 (Mercken et al. 1992) which includes phosphorylated serines around residue 200 (Biernat et al. 1992; Szendrei et al. 1993; Goedert et al. 1993). This epitope is roughly complementary to the epitope of the widely used antibody TAU1 (Binder et al. 1985; Kosik et al. 1988), which recognizes the same region, but only in an unphosphorylated form. These two antibodies can therefore be used as positive and negative indicators of abnormal phosphorylation.

Using a similar approach, the Ser-Pro motifs in recombinant tau have now been found to be phosphorylatable by the kinase activity from brain, as well as several other residues (Gustke et al. 1992). Most sites are clustered in the vicinity of residue 200 or 400, i.e., flanking the region of internal repeats.

The phosphorylation causes a gel shift in several characteristic stages (Fig. 2). Starting from the original unphosphorylated state, we distinguish three main stages with several substages. The PHF-like antibody reaction is detectable from stage 2 onwards. Conversely, other antibodies reacting with normal tau lose their reactivity during stage 2. By these criteria, tau acquires its Alzheimer-like state during stage 2 of the phosphorylation (Lichtenberg-Kraag et al. 1992).

2.2 *Kinases Affecting Tau Protein*

The above results suggest that abnormal phosphorylation of tau occurs predominantly at Ser-Pro motifs (judging by phosphopeptide sequencing and PHF-specific antibodies), and that the corresponding kinase(s) are present in normal brain tissue. This meant that we had to search for a proline directed kinase in brain tissue; it also explained why our earlier searches had failed (e.g., Steiner et al. 1990) since PKA, PKC, casein kinase, and CaM kinase are not proline-directed kinases (see also Correas et al. 1992; Scott et al. 1993). A number of proline-directed kinases are known (reviewed by Hunter 1991). We initially tested the cell cycle kinase p34 (cdc2), which can be complexed with different regulatory subunits (cyclins), but in our hands this cause a neither appreciable phosphorylation, nor the response with PHF-specific antibodies (Drewes et al. 1992). Our next attempt was MAP kinase. This enzyme was prepared from porcine brain and met the criteria required for abnormal phosphorylation: it induced the antibody reaction with PHF-specific antibodies, it incorporated between 12–15 phosphates into the tau molecule, and it phosphorylated Ser-Pro as well Thr-Pro motifs in all recombinant isoforms tested (see Fig. 3). Moreover, when PHFs from Alzheimer brains were dephosphorylated with alkaline phosphatase, they lost their antibody reaction, but regained it when rephosphorylated with MAP kinase.

These data would appear to be compatible with a model in which MAP kinase would be the main cause for abnormal phosphorylation of tau. However, a further search shows that there are still other kinases in the brain extract. One of them is GSK-3, a kinase well known for its role in glycogen metabolism and activation of transcription factors (for review, see Woodgett 1991). When tau is phosphorylated with purified GSK-3, it shows the characteristic M_r shift in the SDS gel, and it acquires antibody reactivities similar to Alzheimer tau (similar to those shown in Fig. 2 for the brain extract kinase activity). However, the degree of phosphorylation is lower than with MAP kinase, only about three to four phosphates per tau molecule. The explanation is that Ser-Pro motifs that regulate the binding of PHF-specific antibodies (Fig. 3) were phosphorylated with similar kinetics and efficiency (Mandelkow et al. 1992; Hanger et al. 1992).

Finally, a search for further kinases showed that certain members of another class of kinases, the cdks (cyclin-dependent kinases), can induce abnormal phosphorylation as well. This includes the kinases cdk2 and cdk5 alias nclk (Baumann et al. 1993). The reaction with cdk2 may not be of physiological relevance because this kinase does not occur in brain tissue (Meyerson et al. 1992). However, cdk5 is abundant in brain and seems to play a role in neuronal development (Lew et al. 1992; Tsai et al. 1993; Shetty et al. 1993). In addition, other workers have described proline-directed kinases phosphorylating tau in similar ways. Examples are a p40 neurofilament kinase which is similar to ERK2, one of the isoforms of MAP kinase (Roder et al. 1993); proline-directed protein kinases (PDPK) which are members of the cdk family (Vulliet et al. 1992; Paudel et al. 1993); and two kinases originally termed tau kinases

I and II (TK-I, TK-II) which now have been identified as GSK-3β (Ishiguro et al. 1993) and a cdk-like kinase (Hisanaga et al. 1993), respectively.

In summary, there are now at least three types of proline-directed kinases which can transform tau into abnormal state, represented by MAP kinase, GSK-3, and cdk 5 (Fig. 4, upper left). They phosphorylate tau to different extents, but since most diagnostic antibodies recognize certain Ser-Pro motifs, the antibody reactivites are similar. It may be significant that these kinases are all involved in cellular signal transduction pathways. This suggests that an abnormal regulation of signal transduction may be important in the generation of the disease. Moreover, MAP kinase, GSK-3, and cdk5 appear to be physically associated with microtubules, as well as with PHFs from Alzheimer brain tissue. Thus any dysregulation of these kinases could affect tau protein directly. There is now growing evidence that some of these kinases may be unusually active during neuronal development so that fetal tau shows similar phosphorylation characteristics as AD tau and reacts with similar diagnostic antibodies (Kanemaru et al. 1992; Watanabe et al. 1993; Bramblett et al. 1993). This would lend support to the idea that in AD the neurons revert to a "fetal-like" state, trying to recover from some stress or toxic effect; such effects are known to activate signal transduction pathways involving MAP kinase.

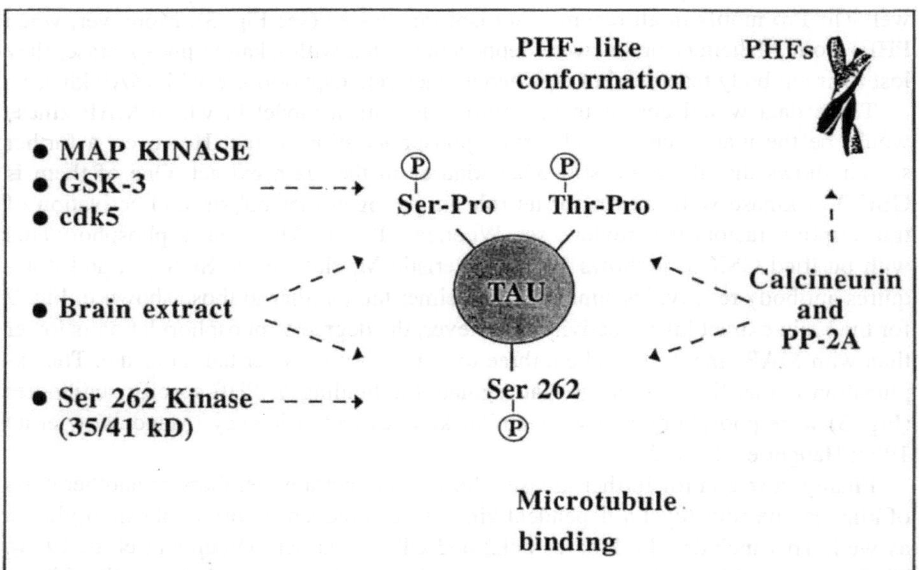

Fig. 4. Kinases and phosphatases acting on tau protein. *Center* Tau contains two classes of phosphorylation sites that appear to be important for Alzheimer's disease. One class comprises the Ser-Pro and Thr-Pro motifs that determine the reactivity with diagnostic antibodies, the second class contains (at least) Ser262 and has a strong effect on microtubule binding. *Left* There are corresponding classes of kinases that phosphorylate tau at these sites: MAP kinase, GSK-3, cdk5 phosphorylate Ser-Pro or Thr-Pro motifs, the Ser262 kinase phosphorylates Ser262 in the first repeat and the corresponding serines in the other repeats. *Right* The phosphatases calcineurin and PP-2A can remove all of these phosphates on tau

2.3 Relationship Between Tau Phosphorylation and Microtubule Binding

It is generally believed that the phosphorylation of tau protein in Alzheimer's disease reduces its affinity to microtubules which therefore break down so that cytoplasmic traffic becomes interrupted. There is substantial evidence that tau from PHFs is abnormally phosphorylated and not bound to microtubules (Lee et al. 1991; Bramblett et al. 1993; Yoshida and Ihara 1993; Lu and Wood 1993; Köpke et al. 1993), but it is less obvious what type of phosphorylation plays a role in the detachment of tau from microtubules. It was therefore of interest to see if any of the kinases described so far

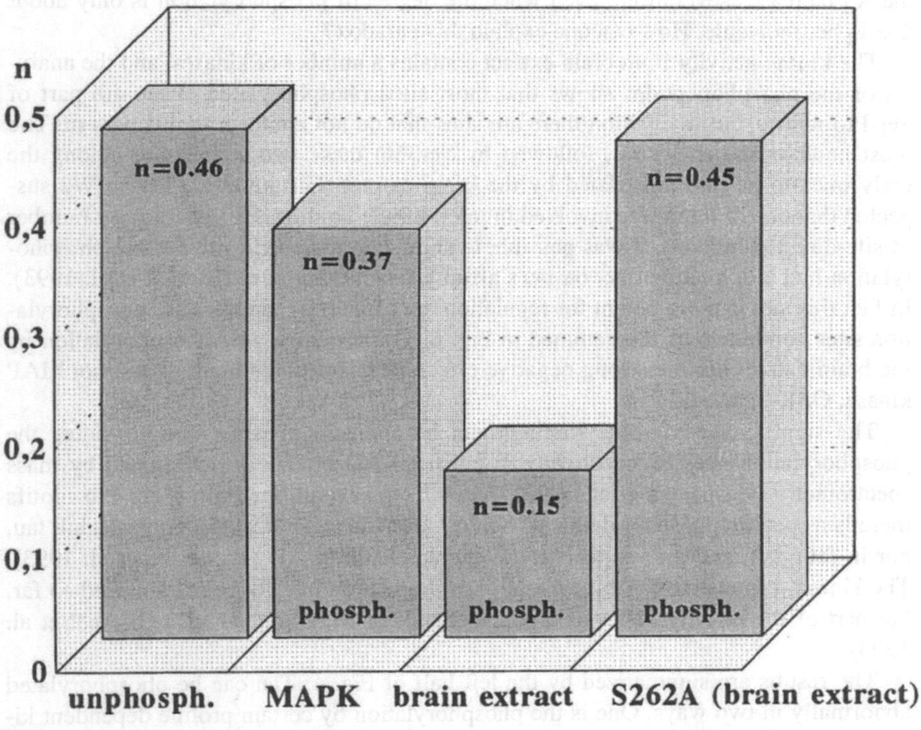

Fig. 5. Summary of binding experiment of tau to microtubules with or without phosphorylation by MAP kinase or the kinase activity from brain extract. *First column* Unphosphorylated tau binds with a stoichiometry of n = 0,46 (tau:tubulin dimer) and a dissociation constant of 1.1 µM (not shown here). *Second column* Phosphorylation by MAP kinase decreases the affinity ≈ two fold ($K_d \approx 1.9$ µM) and reduces the stoichiometry to 0.37. *Third column* Phosphorylation by the brain extract activity strongly reduces both the affinity ($K_d \approx 9.6$ µM) and the stoichiometry (0.15). *Fourth column* When Ser262 is mutated to Ala the phosphorylation dependence disappears (n = 0.45, similar to first column). Note that although MAP kinase is much more efficient in terms of overall phosphorylation, the phosphorylation of Ser262 by the brain extract activity has a much more pronounced effect than all MAP kinase sites combined

had an effect on the binding of tau to microtubules in vitro (Biernat et al. 1993). In a binding experiment, tau constructs are titrated against microtubules stabilized by the drug taxol. This yields two parameters, the stoichiometry and the dissociation constant, where 50% of the tau is bound and 50% is free. Most tau constructs bind with a stoichiometry close to $n = 0.05$ (one tau per tubulin dimers) and K_d values in the μM range. Tau can be highly phosphorylated by MAP kinase which incorporates up to 15 phosphates into tau, i.e., phosphorylated nearly all of the 17 Ser-Pro or Thr-Pro motifs in the largest human isoform. Surprisingly, however, this degree of phosphorylation has only a moderate effect on tau's binding to microtubules (≈ 20% decrease in stoichiometry, Fig. 5). On the other hand, if one phosphorylates tau with the brain extract the stoichiometry drops steeply down to 1/3 of the original value (to $n \approx 0.15$), and the K_d increases severalfold, even when the degree of phosphorylation is only about 2–3 P_i per molecule. How can one explain this paradox?

The kinase activity from brain extract contains a number of kinases, and the analysis of the phosphopeptides shows that most sited phosphorylated in tau are part of Ser-Pro motifs, but in addition there are sites that do not conform to this pattern. The most notable one is Ser262, followed by Ser356; these two serines are among the early ones to be phosphorylated by the brain extract (Gustke et al. 1992). We suspected that one of them was involved in microtubule binding. By generating a number of site-directed mutants, it was possible to show that the single site Ser262 phosphorylation had a dramatic effect on tau's affinity for microtubules (Biernat et al. 1993). In fact this site is more potent for regulation tau's binding than all other phosphorylation sites combined, as diagrammed in Fig. 6. This explains why phosphorylation by the brain extract has the strong negative effect on microtubule binding, but not MAP kinase, GSK-3, or cdk5.

The significance of these findings can be appreciated when one considers the phosphorylation sites found directly in tau from AD brains, as determined by mass spectrometry (Hasegawa et al. 1992). Apart from several Ser-Pro or Thr-Pro motifs there is a specific phosphorylation of Ser262. This does not occur in normal adult tau, nor in fetal tau, and thus appears to be characteristic of AD (Watanabe et al. 1993). The kinase responsible for this type of phosphorylation has not been identified so far, but part of the activity elutes as a pair of bands of M_r35 and 41 kDa (Biernat et al. 1993).

The results are summarized by the left half of Fig. 4: Tau can be phosphorylated abnormally in two ways. One is the phosphorylation by certain proline dependent kinases; this conveys upon tau an altered M_r in SDS gels and an altered response with diagnostic antibodies that distinguish normal tau from PHF tau. The second is the phosphorylation at Ser262 which affects microtubule binding. Not included in this diagram are the other kinases which phosphorylate tau in vitro but show no clear changes with respect to the diagnostic antibodies or microtubule binding; this includes PKA, PKC, CaM kinase, or casein kinase (Baudier and Cole 1987; Steiner et al. 1990; Correas et al. 1992; Scott et al. 1993).

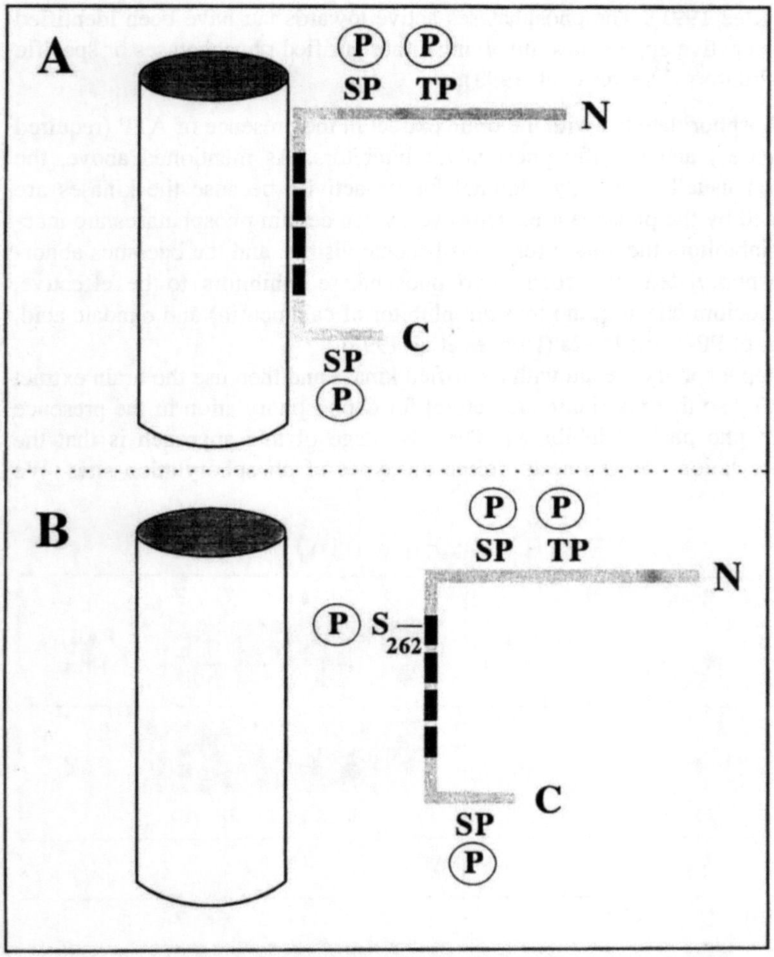

Fig. 6 A, B. Model showing the effect of the phosphorylation of Ser262 on the binding of tau to microtubules. The phosphorylation of the repeat domain at Ser262 has a strong effect on the binding, even though the repeats themselves bind only weakly. The phosphorylation of the Ser-Pro motifs in the flanking regions has only a modest effect on binding, even though these domains mediate the strong binding of tau to microtubules

2.4 Phosphatases Affecting Tau Protein

If there are kinases capable of including abnormal phosphorylation, then there must be phosphatases that counteract the kinases. The kinases must be active at least transiently as part of their role in signal transduction. Since tau is phosphorylated at Ser or Thr residues, the phosphatases must belong to the class of Ser/Thr phosphatases. They are generally classified as PP-1, PP-2a, PP-2b or PP-2c (for review see Cohen 1991); PP-2b is also called calcineurin because of its abundance in brain (reviewed by

Stemmer and Klee 1991). The phosphatases active towards tau have been identified by positive or negative approaches, involving either purified phosphatases or specific phosphatase inhibitors (Drewes et al. 1993).

1. One can phosphorylate tau with the brain extract in the presence of ATP (required for the kinases) and specific phosphatase inhibitors. As mentioned above, the brain extract usually has only minimal kinase activity because the kinases are overwhelmed by the phosphatases. However, when certain phosphatases are inactivated by inhibitors the kinase reactions become visible, and tau becomes abnormally phosphorylated. We found two phosphatase inhibitors to be effective, EGTA (a calcium chelator and thus an inhibitor of calcineurin) and okadaic acid, an inhibitor of PP-1 and PP-2a (Drewes et al. 1993).
2. One can prephosphorylate tau with a purified kinase and then use the brain extract (without ATP so that no kinase are active) for dephosphorylation in the presence of different phosphatase inhibitors. The advantage of this approach is that the prephosphorylation allows one to define the types of phosphorylation sites. We

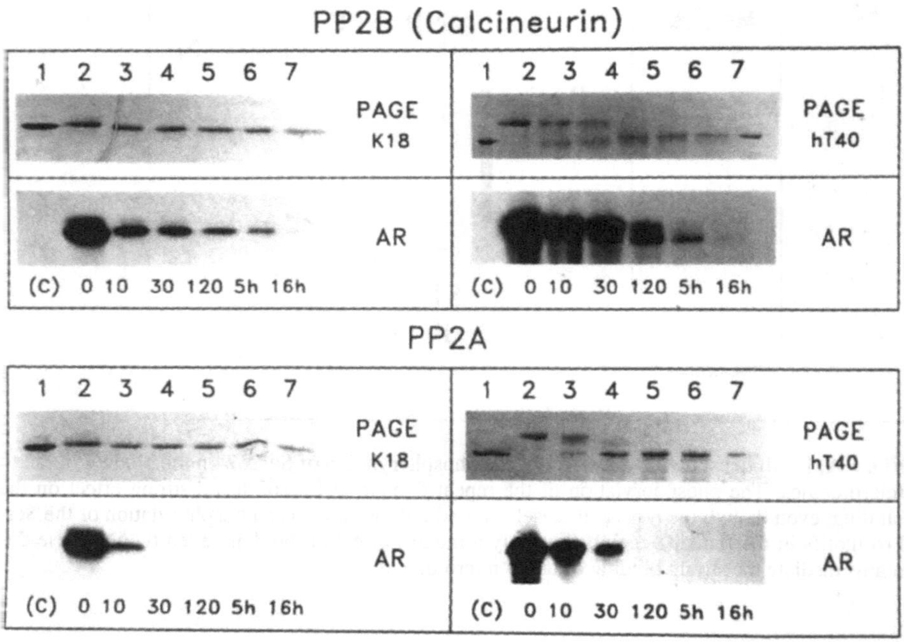

Fig. 7. Dephosphorylation of tau by phosphatases calcineurin (PP-2B, *top half*) and PP-2A (*bottom half*) after prior phosphorylation with the kinase activity from brain extract (Drewes et al. 1993). Full length tau (htau40, *right half*) and K18, a tau construct containing only the repeat region (*left half*), were phosphorylated with extract. The resulting ^{32}P-labeled htau40 (containing 9.0 mol of phosphate per mol) or ^{32}P-labeled K18 (containing 2.5 mol of phosphate per mol) were incubated with PP-2B and PP-2A. *Lane 1* Protein before phosphorylation; *lanes 2–7* after phosphorylation with brain extract during the indicated time periods. The autoradiographs (*AR*) show that tau is removed from both proteins, and by both phosphatases. Note the downward shift in the SDS gel after dephosphorylated of the full length tau *(right half)*

used purified MAP kinase because it phosphorylates tau to a high extent (almost every phosphorylatable motif, see above). The phosphatases present in the brain extract were capable of removing all phosphates from tau, but, as in the previous case, they could be inhibited by EGTA and okadaic acid.

3. One can prephosphorylate tau by purified MAP kinase, and dephosphorylate it with purified phosphatases calcineurin and different forms of PP-2A or PP-1. Calcineurin and PP-2a were both capable of removing all phosphates from pre-phosphorylated tau, PP-1 was not (Fig. 7). This agrees with the inhibitor studies in the first two types of experiment. (We note in passing that okadaic acid inhibits PP-2a and PP-1 with different inhibition constants, ≈ 0.1 and 10 nM, respectively. Thus, in principle, PP-2a should be inhibitable by nM OA while PP-1 would require μM OA; in practice the concentration of PP-2A in cells approaches the μM range so that several μM OA is required to saturate it. This effect can mask the intrinsic difference between the two phosphatases; for a discussion see Goris et al. 1989).

These methods can be applied to other phosphorylation sites; in particular we probed the phosphatases that remove phosphate from Ser262. This was again achieved both by calcineurin and PP-2a (Biernat et al. 1993; Drewes et al. 1993). The results are summarized in the right half of Fig. 4: there are two classes of phosphorylation sites with different effects on tau's properties (Ser/Thr-Pro and Ser262), and there are two types of phosphatases, calcineurin and PP-2a, both of which are active towards all phosphorylation sites of tau.

2.5 Aggregation of Tau into Paired Helical Filaments

Aggregation of tau is probably one of the most crucial problems to be solved because the deposition of PHFs in select neurons is one of the earliest signs of the disease. These PHFs are highly insoluble which makes the extraction and analysis difficult, and they occur only in human AD brains which means that the supply is limited to autopsies. The analysis of Alzheimer tau has improved as methods of isolation have been refined. One can distinguish several levels of pathological aggregation, all of which are based on paired helical filaments. The higher aggregates are neurofibrillary tangles or neuropil threads which consist of coalesced PHFs with other material attached to it (from proteins to metal compounds including aluminium). They are visible by light microscopy after silver staining. With regard to PHFs, one distinguishes between the SDS-insoluble and the SDS-soluble fraction (Wischik et al. 1988; Greenberg and Davies 1990). Both are more highly phosphorylated (6–8 P_i per tau) than normal adult tau (≈ 2 P_i, Ksiezak-Reding et al. 1992). The SDS-insoluble fraction probably represents a later and more extensive state of aggregation, it is more highly ubiquitinated, and it is resistant to proteolysis (pronase removes a fuzzy coat but leaves the core aggregated, Kondo et al. 1988; Wischik et al. 1988; Kziezak-Reding and Wall 1994). The SDS-soluble fraction can also be digested by pronase; analysis of this fraction has shown that PHFs consist mainly of tau, and moreover all isoforms of tau (Brion et al. 1991; Goedert et al. 1992). The states of lower aggregation are more difficult to distinguish. Obviously, even Alzheimer brains

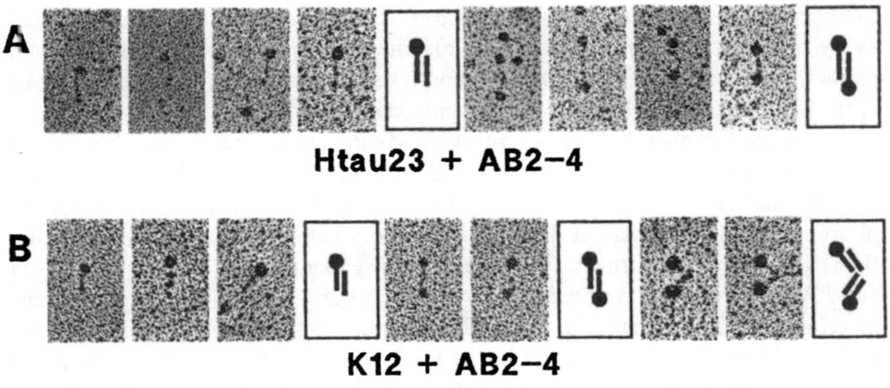

Fig. 8 A, B. Electron micrographs of tau isoform htau23 or construct K12 (Wille et al. 1992). The rod-like particles are about 35 nm *(htau23)* (**A**) or 25 nm *(K12)* (**B**) long (Wille et al. 1992). They are labeled at one or both ends with antibody *2–4*, showing that the particles can form antiparallel dimers and more complex structures (see interpretative diagrams)

contain a fraction of normal tau which is highly soluble and in a state of low phosphorylation. However, there is also incipient aggregation in the form of soluble AD tau (Köpke et al. 1993).

The mechanism of PHF aggregation can be analyzed by inducing the process with recombinant tau constructs, studying the structures formed and comparing them with PHFs from Alzheimer brains. The advantage is that the starting material is soluble, it has a defined sequence and composition, and it can be modified by phosphorylation. Using this approach, we discovered several features of tau (Wille et al. 1992). (1) Tau is a highly extended molecule, with lengths around 35–50 nm (depending on isoform). (2) Tau can associate into dimers which are antiparallel and roughly in register (Fig. 8). This can be shown by immuno-EM using antibody labels that bind to the end

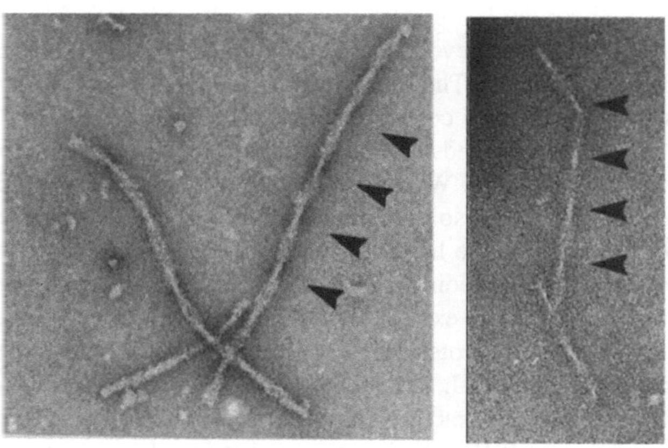

Fig. 9. Synthetic paired helical filaments assembled from tau construct K12 (consisting of repeats 1, 3, 4, and a short tail. The crossover periodicity is about 75–80 nm *(arrowheads)* (Wille et al. 1992)

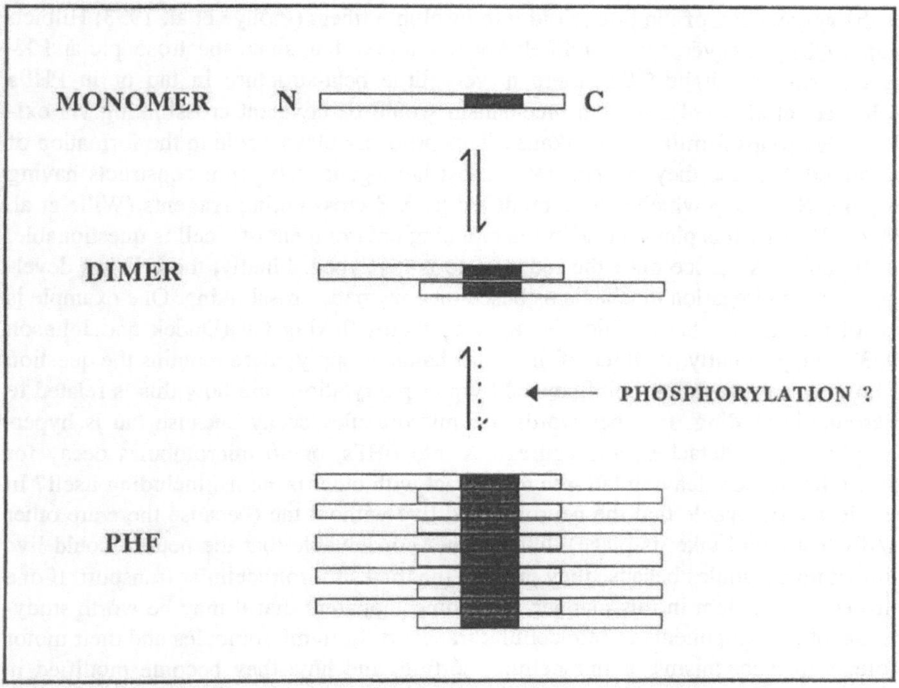

Fig. 10. Hypothetical steps in PHF assembly. Tau monomers aggregate into antiparallel dimers through their repeat domains. The dimers assemble into PHFs

of the tau molecule. (3) Certain tau constructs (comprising mainly the repeat region of tau) and their dimers can associate into paired helical filaments which are very similar to those found in AD (Fig. 9). Such filaments are also formed from chemically crosslinked dimers of the repeat region of tau, suggesting that the dimer stage is an intermediate level of tau aggregation, as diagrammed in Fig. 10. Whether and how the aggregation of tau into synthetic PHFs depends on phosphorylation is not clear at present.

The formation of synthetic PHFs from recombinant tau was an important step towards elucidating the assembly mechanism, but it also illustrates our lack of detailed understanding of the control mechanisms. It is not yet possible to cycle between the states of assembly and disassembly (as is possible with other self-assembling polymers such as actin filaments or microtubules). It is also not yet clear what the main principle behind the assembly is. Insoluble protein is commonly formed from denatured protein when the hydrophobic interior is exposed and then coalesces with other molecules. Such hydrophobic interactions are difficult to envisage with tau protein because its composition is unusually hydrophilic (Lee et al. 1988); this probably explains why it is heat-stable and resistant to acid treatment (Cleveland and Kirschner 1977; Fellous et al. 1977). Another possibility is a high extent of hydrogen-bonded β-structure (as proposed for "self-complementary" peptides by Zhang et al. 1993). Extensive β-structure indeed makes proteins very stable, and this is probably the basis for the aggregation of the β-amyloid into insoluble fibers (Inouye et al. 1993; Hilbich et al. 1992). However, this is unlikely for the case of tau, since spectroscopic and X-ray experiments suggest that there is very little beta-structure in tau or in PHFs (Schweers et al. 1994). A third mechanism would be covalent crosslinking via oxidized SH groups forming S-S linkages. This probably plays a role in the formation of tau dimers because they are generated most homogeneously from constructs having only one SH group which can be modified by S-S-crosslinking reagents (Wille et al. 1992). Whether this plays a role in the reducing environment of a cell is questionable, but it could take place once the redox state is perturbed. Finally, the cell has developed other aggregation mechanisms based on enzymatic crosslinking. One example is that of transglutaminase, which is capable of crosslinking tau (Dudek and Johnson 1993). Independently of which of these mechanisms apply, there remains the question of how tau's aggregation is influenced by phosphorylation, and how this is related to microtubule binding. In other words: do microtubules decay because tau is hyperphosphorylated, detaches, and aggregates into PHFs, or do microtubules decay for some other reason, leaving tau free to interact with other proteins, including itself? In fact, it is conceivable that the neuron could live without tau (because there are other MAPs that could take its place), but it is not conceivable that the neuron could live without microtubules because they provide the tracks for intracellular transport. If one phrases the problem in this manner, it becomes apparent that it may be worth studying the other components of intracellular transport, from microtubules and their motor proteins to mechanisms of intracellular sorting, and how they become modified in Alzheimer's disease.

Acknowledgments. We would like to thank L. Binder, A. Vandevoorde, and M. Mercken for antibodies, and M. Goedert for the clones of human tau. Phosphopeptide sequencing was done in collaboration with Dr. H. E. Meyer (Univ. Bochum). Brain tissue was generously provided by the Bryan Alzheimer Disease Research Center (Duke University Medical Center, Durham, NC), the Brain Tissue Resource Center (McLean Hospital/Harvard Medical School, Belmont, MA), and the Alzheimer Research Center (Univ. Rochester Medical School, Rochester, NY). This work was supported by Bundesministerium für Forschung und Technologie (BMFT) and the Deutsche Forschungsgemeinschaft (DFG).

References

Anderton BH (1993) Expression and processing of pathological proteins in Alzheimer's disease. Hippocampus 3:227–237

Bancher C, Grundke-Iqbal I, Iqbal K, Fried V, Smith H and Wisniewski H (1991) Abnormal phosphorylation of tau precedes ubiquitination in neurofibrillary pathology of Alzheimer disease. Brain Research 539:11–18

Baudier J and Cole RD (1987) Phosphorylation of tau proteins to a state like that in Alzheimer's brain is catalyzed by a calcium/calmodulin-dependent kinase and modulated by phospholipids. J Biol Chem 262:17577–17583

Baumann K, Mandelkow E-M, Biernat J, Piwnica-Worms H, Mandelkow E (1993) Abnormal Alzheimer-like phosphorylation of tau protein by cyclin-dependent kinases cdk2 and cdk5. FEBS Lett 336:417–242

Bialojan C and Takai A (1988) Inhibitory effect of a marine sponge toxin, okadaic acid, on protein phosphatases. Biochem J 256:283–290

Biernat J, Mandelkow E-M, Schröter C, Lichtenberg-Kraag B, Steiner B, Berling B, Meyer HE, Mercken M, Vandermeeren A, Goedert M, Mandelkow E (1992) The switch of tau protein to an Alzheimer-like state includes the phosphorylation of two serine-proline motifs upstream of the microtubule binding region. EMBO J 11:1593–1597

Biernat J, Gustke N, Drewes G, Mandelkow E-M, Mandelkow E (1993) Phosphorylation of serine 262 strongly reduces the binding of tau protein to microtubules: Distinction between PHF-like immunoreactivity and microtubule binding. Neuron 11:153–163

Binder LI, Frankfurter A and Rebhun L (1985) The distribution of tau in the mammalian central nervous system. J Cell Biol 101:1371–1378

Braak H, and Braak E (1991) Neuropathological staging of Alzheimer-related changes. Acta Neuropath 82:239–259

Braak H and Braak E (1994) Pathology of Alzheimer's disease. In: Neurodegenerative diseases, Eds. D Calne et al., WB Saunders Co., Philadelphia, pp 585–613

Braak E, Braak H and Mandelkow E-M (1994) A sequence of cytoskeleton changes related to the formation of neurofibrillary tangles and neuropil threads. Acta Neuropath 87:554–567

Bramblett GT, Goedert M, Jakes R, Merrick SE, Trojanowski JQ and Lee VMY (1993) Abnormal tau phosphorylation at Ser(396) in Alzheimer's disease recapitulates development and contributes to reduced microtubule binding. Neuron 10:1089–1099

Brandt R and Lee G (1993) Functional organization of microtubule-associated protein tau: identification of regions which affect microtubule growth, nucleation, and bundle formation in vitro. J Biol Chem 268:3414–3419

Brion J, Passareiro H, Nunez J and Flament-Durand J (1985) Mise en évidence immunologique de la proteine tau au niveau des lesions de dégénérescence neurofibrillaire de la maladie d'Alzheimer. Arch Biol 95:229–235

Brion JP, Hanger DP, Couck AM and Anderton BH (1991) A68 proteins in Alzheimers disease are composed of several tau isoforms in a phosphorylated state which affects their electrophoretic mobilities. Biochem J 279:831–836

Butner KA and Kirschner MW (1991) Tau-protein binds to microtubules through a flexible array of distributed weak sites. J Cell Biol 115:717–730

Cleveland DW, Hwo S-Y and Kirschner MW (1977a) Physical and chemical properties of purified tau factor and the role of tau in microtubule assembly. J Mol Biol 116:227–247

Cohen P (1991) Classification of protein serine/threonine phosphatases: identification and quantitation in cell extracts. Meth Enzym 201:389–398

Correas I, Diaznido J and Avila J (1992) Microtubule associated protein tau is phosphorylated by protein kinase C on its tubulin binding domain. J Biol Chem 267:15721–15728

Couchie D, Mavilia C, Georgieff I, Liem R, Shelanski M and Nunez J (1992) Primary structure of high molecular weight tau present in the peripheral nervous system. Proc Natl Acad Sci USA 89:4378–4381

Crowther RA (1991) Straight and paired helical filaments in Alzheimer disease have a common structural unit. Proc Natl Acad Sci USA 88:2288–2292

Crowther RA, Olesen OF, Smith MJ, Jakes R, Goedert M (1994) Assembly of Alzheimer-like filaments from full-length tau protein. FEBS Letters 337:134–138

Drewes G, Lichtenberg-Kraag B, Döring F, Mandelkow EM, Biernat J, Goris J, Doree M and Mandelkow E (1992) Mitogen-activated protein (MAP) kinase transforms tau protein into an Alzheimer-like state. EMBO J 11:2131–2138

Drewes G, Mandelkow E-M, Baumann K, Goris J, Merlevede W, Mandelkow E (1993) Dephosphorylation of tau protein and Alzheimer-paired helical filaments by calcineurin and phosphatase-2A. FEBS Lett 336:425–432

Dudek SM and Johnson GVW (1993) Transglutaminase catalyzes the formation of SDS-insoluble, Alz50-reactive polymers of tau. J Neurochem 61:1159–1162

Ennulat DJ, Liem RKH, Hashim GA and Shelanski ML (1989) Two separate 18-amino acid domains of tau promote the polymerization of tubulin. J Biol Chem 264:5327–5330

Fellous A, Francon J, Lennon AM and Nunez J (1977) Microtubule assembly in vitro: purification of assembly promoting factors. Eur J Biochem 78:167–174

Goedert M, Wischik C, Crowther R, Walker J and Klug A (1988). Cloning and sequencing of the cDNA encoding a core protein of the paired helical filament of Alzheimer disease: identification as the microtubule-associated protein tau. Proc Natl Acad Sci USA 85:4051–4055

Goedert M, Spillantini M, Jakes R, Rutherford D and Crowther RA (1989b) Multiple isoforms of human microtubule-associated protein-tau: sequences and localization in neurofibrillary tangles of Alzheimer's disease. Neuron 3:519–526

Goedert M, Spillantini G, Cairns NJ and Crowther RA (1992) Tau proteins of Alzheimer paired helical filaments: abnormal phosphorylation of all six brain isoforms. Neuron 8:159–168

Goedert M, Jakes R, Crowther RA, Six J, Lübke U, Vandermeeren M, Cras P, Trojanowski JQ and Lee VMY (1993) The abnormal phosphorylation of tau protein at Ser202 in Alzheimer's disease recapitulates phosphorylation during development. Proc Natl Acad Sci USA 90:5066–5070

Goedert M (1993) Tau protein and the neurofibrillary pathology of Alzheimer's disease. Trends in Neurosci 16:460–465

Goris J, Hermann J, Hendrix P, Ozon R, Merlevede W (1989) Okadaic acid, a specific protein phosphatase inhibitor, induces maturation and MPF formation in *Xenopus laevis* oocytes. FEBS Letters 245:91–94

Greenberg SG and Davies P (1990) A preparation of Alzheimer-paired helical filaments that displays distinct tau-proteins by polyacrylamide-gel electrophoresis. Proc Natl Acad Sci USA 87:5827–5831

Grundke-Iqbal I, Iqbal K, Tung Y, Quinlan M, Wisniewski H, Binder L (1986a) Abnormal phosphorylation of the microtubule-associated protein tau in Alzheimer cytoskeletal pathology. Proc Natl Acad Sci USA 83:4913–4917

Gustke A, Steiner B, Mandelkow E-M, Biernat J, Meyer HE, Goedert M, Mandelkow E (1992) The Alzheimer-like phosphorylation of tau protein reduces microtubule binding and involves Ser-Pro and Thr-Pro motifs. FEBS Letters 307:199–205

Gustke N, Trinczek B, Biernat J, Mandelkow E-M, Mandelkow E (1994) Domains of tau protein and interactions with microtubules. Biochemistry 33:9511–9522

Hanger D, Hughes K, Woodgett J, Brion J, Anderton B (1992) Glycogen-synthase kinase-3 induces Alzheimer's disease-like phosphorylation fo tau: generation of paired helical filament epitopes and neuronal localization of the kinase. Neurosci Lett 147:58–62

Hasegawa M, Morishima-Kawashima M, Takio K, Suzuki M, Titani K, Ihara Y (1992) Protein sequence and mass spectrometric analyses of tau in the Alzheimer's disease brain. J Biol Chem 26:17047–17054

Hilbich C, Kisters-Woike B, Reed J, Masters C and Beyreuther K (1992) Substitutions of hydrophobic amino-acids reduce the amyloidogenicity of Alzheimer's disease βA4 peptides. J Mol Biol 228:460–473

Himmler A, Drechsel D, Kirschner M and Martin D (1989) Tau consists of a set of proteins with repeated C-terminal microtubule-binding domains and variable N-terminal domains. Molec Cell Biol 9:1381–1388

Hisanaga S, Ishiguro K, Uchida T, Okumura E, Okano T, Kishimoto T (1993) Tau-protein kinase II has a similar characteristic to cdc2 kinase for phosphorylating neurofilament proteins. J Biol Chem 268:15056–15060

Hunter T (1991) Protein kinase classification. Meth Enzym 200:3–37

Inouye H, Fraser PE and Kirschner DA (1993) Structure of beta-crystallite assemblies formed by Alzheimer beta-amyloid protein analogues: analysis by X-ray diffraction. Biophys J 64:502–519

Ishiguro K, Shiratsuchi A, Sato S, Omori A, Arioka M, Kobayashi S, Uchida T and Imahori K (1993) Glycogen-synthase kinase 3-beta is identical to tau protein kinase I generating several epitopes of paired helical filaments. FEBS Letters 325:167–172

Kanai Y, Chen J, Hirokawa N (1992) Microtubule bundling by tau proteins in vivo: analysis of functional domains. EMBO J 11:3953–3961

Kanemaru K, Takio K, Miura R, Titani K and Ihara Y (1992) Fetal-type phosphorylation of the tau in paired helical filaments. J Neurochem 58:1667–1675

Knops J, Kosik K, Lee G, Pardee J, Cohengould L, McConlogue L (1991) Overexpression of tau in a nonneuronal cell induces long cellular processes. J Cell Biol 114:725–733

Kondo J, Honda T, Mori H, Hamada Y, Miura R, Ogawara M and Ihara Y (1988) The carboxyl third of tau is tightly bound to paired helical filaments. Neuron 1:827–834

Köpke E, Tung Y, Shaikh S, Alonso A, Iqbal K, Grundke-Iqbal I (1993) Microtubule-associated protein tau: abnormal phosphorylation of a non-paired helical filament pool in Alzheimer's disease. J Biol Chem 268:24374–24384

Kosik K, Orecchio L, Binder L, Trojanowski J, Lee V and Lee G (1988) Epitopes that span the tau molecule are shared with paired helical filaments. Neuron 1:817–825

Ksiezak-Reding H and Yen SH (1991) Structural stability of paired helical filaments requires microtubule-binding domains of tau: a model for self-association. Neuron 6:717–728

Ksiezak-Reding H, Liu WK and Yen SH (1992) Phosphate analysis and dephosphorylation of modified tau associated with paired helical filaments. Brain Res 597:209–219

Ksiezak-Reding H and Wall JS (1994) Mass and physical dimensions of 2 distinct populations of paired helical filaments. Neurobiol Aging 15:11–19

Lee G, Cowan N and Kirschner M (1988) The primary structure and heterogeneity of tau protein from mouse brain. Science 239:285 11288

Lee VMY, Balin BJ, Otvos L and Trojanowski JQ (1991) A68 – a major subunit of paired helical filaments and derivatized forms of normal tau. Science (Wash) 251:675–678

Lew J, Winkfein RJ, Paudel HK and Wang JH (1992b) Brain proline-directed protein kinase is a neurofilament kinase which displays high sequence homology to p34 (cdc2). J Biol Chem 267:25922–25926

Lichtenberg-Kraag B, Mandelkow E-M, Biernat J, Steiner B, Schröter C, Gustke N, Meyer HE, Mandelkow E (1992). Phosphorylation dependent interaction of neurofilament antibodies with tau protein: epitopes, phosphorylation sites, and relationship with Alzheimer tau. Proc Natl Acad Sci USA 89:5384–5388

Lo MMS, Fieles AW, Norris TE, Dargis PG, Caputo CB, Scott CW, Lee VMY, Goedert M (1993) Human tau isoforms confer distinct morphological and functional properties to stably transfected fibroblasts

Lu Q, Wood JG (1993) Functional studies of Alzheimer's disease tau protein. J Neurosci 13:508–515

Mandelkow E-M, Drewes G, Biernat J, Gustke N, Can Lint J, Vandenheede JR and Mandelkow E (1992) Glycogen synthase kinase-3 and the Alzheimer-like state of microtubule-associated protein tau. FEBS Lett 314:315–321

Mandelkow E-M and Mandelkow E (1993) Tau as a marker of Alzheimer's disease. TIBS 18:480–483

Mercken M, Vandermeeren M, Lübke U, Six J, Boons J, Van de Voorde A, Martin J-J and Gheuens J (1992) Monoclonal antibodies with selective specificity for Alzheimer tau are directed against phosphatase-sensitive epitopes. Acta Neuropathol 84:265–272

Meyer HE, Hoffmann-Posorske E and Heilmeyer LMG (1991) Determination and location of phosphoserine in proteins and peptides by conversion to S-ethyl-cysteine. Meth Enzymol 201:169–185

Meyerson M, Enders GH, Wu CL, Su LK, Gorka C, Nelson C, Harlow E and Tsai LH (1992) A family of human cdc2-related protein-kinases. EMBO J 11:2909–2917

Morishima-Kawashima M, Hasegawa M, Takio K, Suzuki M, Titani K, Ihara Y (1993) Ubiquitin is conjugated with amino-terminally processed tau in paired helical filaments. Neuron 10:151–1160

Novak M, Kabat J and Wischik CM (1993) Molecular characterization of the minimal protease resistant tau-unit of the Alzheimer's disease-paired helical filament. EMBO J 12:365–370

Paudel H, Lew J, Ali Z, Wang J (1993) Brain proline-directed protein kinase phosphorylates tau on sites are abnormally phosphorylated in tau associated with Alzheimer's-paired helical filaments. J Biol Chem 268:23512–23518

Roder HM, Eden PA and Ingram VM (1993) Brain protein kinase pk40(erk) converts tau into a PHF-like form as found in Alzheimer's disease. Biochem Biophys Res Comm 193:639–647

Schweers O, Schönbrunn-Hanebeck E, Marx A, Mandelkow E (1994) Structural studies of tau protein and Alzheimer-paired helical filaments show no evidence for β structure. J Biol Chem 269:24290–24297

Scott C, Spreen R, Herman J, Chow F, Davison M, Young J, Caputo C (1993b) Phosphorylation of recombinant tau by cAMP-dependent protein kinase: identification of phosphorylation sites and effect on microtubule assembly. J Biol Chem 268:1166–1173

Shetty KT, Link WT and Pant HC (1993) Cdc2-like kinase from rat spinal-cord specifically phosphorylates KSPXK motifs in neurofilament proteins: isolation and characterization. Proc Natl Acad Sci USA 90:6844–6848

Steiner B, Mandelkow E-M, Biernat J, Gustke N, Meyer HE, Schmidt B, Mieskes G, Söling HD, Drechsel D, Kirschner MW, Goedert M and Mandelkow E (1990) Phosphorylation of microtubule-associated protein tau: identification of the site for Ca^{2+}-calmodulin-dependent kinase and relationship with tau phosphorylation in Alzheimer tangles. EMBO J 9:3639–3544

Stemmer P and Klee C (1991) Serine/threonine phosphatases in the nervous system. Curr Opin Neurobiol 1:53–64

Szendrei GI, Lee VM-Y and Otvos L (1993) Recognition of the minimal epitope of monoclonal antibody Tau-1 depends upon the presence of a phosphate group but not its location. J Neurosci Res 34:243–249

Tsai LH, Takahashi T, Caviness V, Harlow E (1993) Activity and expression pattern of cyclin-dependent kinase 5 in the embryonic mouse nervous system. Development 119:1029–1040

Vulliet R, Halloran S, Braun R, Smith A and Lee G (1992) Proline-directed phosphorylation of human tau protein. J Biol Chem 267:22570–22574

Watanabe A, Hasegawa M, Suzuki M, Takio K, Morishima-Kawashima M, Titani K, Arai T, Kosik KS and Ihara Y (1993) In-vivo phosphorylation sites in fetal and adult rat tau. J Biol Chem 268:25712–25717

Wille H, Drewes G, Biernat J, Mandelkow E-M, Mandelkow E (1992c) Alzheimer-like paired helical filaments and antiparallel dimers formed from microtubule-associated protein tau in vitro. J Cell Biol 118:573–584

Wischik C, Novak M, Thogersen H, Edwards P, Runswick M, Jakes R, Walker J, Milstein C, Roth M and Klug A (1988a) Isolation of a fragment of tau derived from the core of the paired helical filament of Alzheimer disease. Proc Natl Acad Sci USA 85:4506–4510

Woodgett JR (1991a) A common denominator linking glycogen metabolism, nuclear oncogenes, and development. TIBS 16:177–181

Yoshida H, Ihara Y (1993) Tau in paired helical filaments is functionally distinct from fetal tau: assembly incompetence of paired helical filament tau. Journal of Neurochem 61:1183–1186

Zhang S, Holmes T, Lockshin C and Rich A (1993) Spontanous assembly of a self-complementary oligopeptide to form a stable macroscopic membrane. Proc Natl Acad Sci USA 90:3334–3338

Springer-Verlag and the Environment

We at Springer-Verlag firmly believe that an international science publisher has a special obligation to the environment, and our corporate policies consistently reflect this conviction.

We also expect our business partners – paper mills, printers, packaging manufacturers, etc. – to commit themselves to using environmentally friendly materials and production processes.

The paper in this book is made from low- or no-chlorine pulp and is acid free, in conformance with international standards for paper permanency.

MIX
Papier aus verantwortungsvollen Quellen
Paper from responsible sources
FSC® C105338

If you have any concerns about our products, you can contact us on
ProductSafety@springernature.com

In case Publisher is established outside the EU, the EU authorized representative is:
Springer Nature Customer Service Center GmbH
Europaplatz 3, 69115 Heidelberg, Germany

Printed by Libri Plureos GmbH
in Hamburg, Germany